MINDFUL COGNITIVE BEHAVIORAL THERAPY

◇

*A Simple Path to Healing,
Hope, and Peace*

SETH J. GILLIHAN

HarperOne
An Imprint of HarperCollins*Publishers*

The content of this book is intended to inform, entertain, and provoke your thinking. This is not intended as medical advice. Neither the author nor the publisher can be held responsible or liable for any loss or claim arising from the use, or misuse, of the content of this book.

Names and identifying details of patients have been changed to protect their identity.

MINDFUL COGNITIVE BEHAVIORAL THERAPY. Copyright © 2022 by Seth J. Gillihan. All rights reserved. Printed in the United States of America. No part of this book may be used or reproduced in any manner whatsoever without written permission except in the case of brief quotations embodied in critical articles and reviews. For information, address HarperCollins Publishers, 195 Broadway, New York, NY 10007.

HarperCollins books may be purchased for educational, business, or sales promotional use. For information, please email the Special Markets Department at SPsales@harpercollins.com.

FIRST HARPERCOLLINS PAPERBACK PUBLISHED IN 2024

Designed by Leah Carlson-Stanisic

Library of Congress Cataloging-in-Publication Data is available upon request.

ISBN 978-0-06-307572-6

24 25 26 27 28 LBC 5 4 3 2 1

For Marcia Lynn Leithauser

If you want to discover the truth about God,
don't strive for things that lie beyond you.

Draw your thoughts inward to the center, and
seek to become one and simple in your soul.

Let go of all that distracts you, all you desire,
and come home to yourself, and when you do,

you'll become the truth you first sought.

—Meister Eckhart (1260–1328)[1]

CONTENTS

1

HEAR THE CALL

If there's a common longing among the hundreds of people I've treated in therapy as a clinical psychologist, it's for an end to their pain. But my own journey through depression taught me that relieving symptoms is not enough. More than finding a cure for suffering, our deepest longing is for peace. That distinction is what this book is about.

Most people who come to see me are dealing with some form of overwhelming anxiety: panic, constant worry, obsessive-compulsive disorder (OCD), social fears. Many are healing from trauma, sometimes recent, sometimes from their childhood. Some are fighting through daily depression or chronic illness or are wondering if their marriage can be saved. Others are desperate for a good night's sleep. In one way or another, they're yearning for relief from the stress and strain of life.

People seek out my services because they believe I can help them find relief and peace through *cognitive behavioral therapy* (CBT), the most scientifically tested therapeutic method practiced today. CBT is a straightforward approach that integrates two components:

- **Cognitive therapy** for practicing healthy patterns of thought
- **Behavioral therapy** for helping us choose actions that lead us toward our goals

Treatment tends to be brief, typically eight to fifteen sessions, and addresses current problems rather than focusing on a person's childhood and relationship with their parents. I was drawn to this approach early in my graduate training because I wanted to relieve suffering, and CBT seemed like the most efficient path to healing.

But after a number of years helping others as a CBT therapist, I discovered that I needed help as well. I had slowly dropped into a deep depression, and despite all my training, I was struggling to find my way out. Eventually, as I groped my way forward, I discovered something surprising and significant. I learned that CBT could be more than a means for eliminating symptoms, which was how I had been using it. When combined with mindfulness practices, it could also address questions of meaning, purpose, and even spiritual peace.

That is quite a claim, I know. But be assured this is not one of those books by some self-designated guru who claims to have finally figured out the secret of the universe and wants people to follow behind. I am by no means the first person to walk this road.

My goal, instead, is to simplify the process that I found to be so incredibly helpful so that as many people as possible can experience it for themselves. This life-changing approach can be summarized in three words, making it easy to remember on the fly when you need it: Think Act Be.

Descent

I was motivated to become a psychologist in part by what I knew of my grandfather's emotional struggles and his suicide eight years before I was born. Frank Rollin Gillihan was haunted by horrific memories of naval combat in the South Pacific during World War II; I wondered what his life would have been like if he had gotten effective psychological treatment. Maybe he would have lived to meet his grandchildren. His only child—my father—would have been spared the pain of losing his dad to suicide. That pain was a felt presence throughout my childhood, often apparent in my dad's irritability and temper. At other times it was raw grief, such as when I was eight and walked in on my parents in our laundry room. My mom had her arm around my dad as he sobbed, a stack of old family photos in his hand.

I did my training at the University of Pennsylvania, the birthplace of many CBT treatment programs. The faculty were deeply engaged in developing short-term, effective treatments and testing them in rigorous clinical trials, and my faith in the power of CBT deepened as I witnessed its effects firsthand. I saw the power of thoughts to affect our emotions. I learned how simple changes in our actions can boost mood and bring greater fulfillment.

I continued at Penn after I graduated and took a faculty position in an anxiety research center, where I oversaw a treatment study of CBT for post-traumatic stress disorder (PTSD). Participants came from the surrounding community and the local Veterans Affairs hospital, men and women haunted by traumatic memories of violence and pain. After the twelve-session protocol, many were transformed, freed from nightmares and flashbacks and ready to live their lives again. I often thought of my grandfather.

When I left Penn and opened a solo private practice, I continued to offer CBT. It was exhilarating to see the dramatic effects that a few sessions—sometimes just five or six—could have on a person's life. The grip of anxiety would loosen, depression would lift, sleep would improve. My schedule quickly filled with people who wanted an action-focused way to feel better.

But as I continued to practice, I was often struck by changes my patients were experiencing that seemed to transcend simple symptom reduction. People described feeling lighter, freer, more connected to a version of themselves that they liked. Their family members told me with tears in their eyes that they finally had their loved ones back.

I wasn't sure what to make of these changes, since they didn't fit neatly with my CBT view of therapy that focused on measurable outcomes. At times, I even envied the deep work they were doing and the new levels of peace and happiness that they'd found.

I was especially struck by the deep changes I saw with Paul, a young dad who was out of work.[1] His childhood had been tough, and Paul had hated himself for as long as he could remember. His father had left the family when Paul was five, and Paul always felt that he was his mom's least favorite child. He had battled alcohol addiction early in his life and had struggled in his closest relationships.

Paul's biggest challenge was feeling like a failure to his young daughter and son. He'd been deeply wounded by his dad's departure and had always vowed he'd be a father his children could be proud of. With his job loss and the depression that followed, he thought they must see him as a pathetic disappointment. He choked on the words every time he tried to talk about disappointing his kids but would wave off the tissue I offered. His shame would quickly turn to anger at himself for being a "crybaby" as he roughly brushed away the tears with the heel of his hand. Paul denied being an imminent threat to himself but said he often imagined ending his life.

We had worked together for many months—longer than a textbook course of CBT—and he had made slow but steady progress. He had gradually and consistently started doing more activities that brought him enjoyment and a sense of accomplishment, which dramatically improved his mood. Paul also learned to recognize the lie in his horrible thoughts about himself, such as "I'm worthless" and "everyone would be better off without me." And yet there remained a deeper undercurrent of self-hatred that seemed to resist the efforts he was making in therapy.

But one day Paul shocked me. The tears came, and he let them. Only this time, he wasn't crying about being a lousy father. He was crying for his five-year-old self who had lost his dad and had never known love until he had kids of his own. Through his tears, he told me he was starting to feel love for himself. I was wiping away tears of my own. I had hoped for this day as long as I'd known Paul, who in truth was easy to love.

But when Paul's relationship with himself finally shifted, it caught me by surprise. Our self-directed thoughts and feelings are stubbornly hard to change. I was used to seeing patients make incremental changes in these areas but often somewhat begrudgingly and with a lingering sense of self-loathing. Paul's transformation was of a different kind. It was as if a barrier had been torn down between his heart and himself, releasing a wave of self-love that he had held back for decades. He could finally see that his wounds and suffering called for compassion, not disgust.

Paul didn't just stop hating himself or simply stop being depressed. Paul was transformed. He became the father and husband he'd always wanted to be. How did our therapy help to bring that about? I wasn't sure I knew.

Discovery

It was only later that evening after our breakthrough session that the irony hit me. Just that week, I'd been torn up inside about being a disappointment to my own wife and kids. I'd been struggling with health issues for a couple of years. It started with persistent problems with my voice—laryngitis, a burning sensation in my throat, difficulty making myself heard. I struggled to meet the vocal demands of therapy and the teaching I was doing at a local college. Over time, I was plagued by a long and growing list of nonspecific symptoms: poor sleep, physical exhaustion, mental confusion, body pain, heat intolerance, and digestive problems, among many others. Frequent visits to many specialists and alternative therapists brought few answers and little relief and a growing stack of medical bills.

My world shrank as my struggles continued. I had to stop most forms of exercise because of the fatigue, and I no longer got together with friends, since it was so hard to talk. Even at home I rarely spoke, having exhausted my limited "vocal reserve" by the time I was done with work. I was forced to reduce my clinical hours because of my voice limitations and low energy, which led to serious financial strain for my family.

In hindsight, I realize that depression was nearly inevitable, given my circumstances: chronic stress, social isolation, lack of exercise, poor sleep. I'd witnessed this pattern countless times in my clinical work, and now I was experiencing it myself. It took me awhile to recognize that I had sunk into a deep depression, wanting to die and thinking my family would be better off without me. My wife, Marcia, was incredibly supportive, but she couldn't take away the lows or the self-loathing. She would reassure me when I bottomed out: "Seth, you're doing the best you can. It's not your fault that you're sick." Meanwhile, I was silently shouting in my head, over and over, "I fucking *hate* myself!"

My depression dragged on for months. In its depths I felt lost, bewildered, and alone. I didn't know what had brought me to this place and felt too exhausted and confused to escape from it. I was crying all the time. I cried on my walk to work, having no idea how I would make it through my day. I cried on the way home, struggling up tiny hills and feeling as if I were wearing lead boots. I cried on the couch in my office, where I napped between patients, careful about my head placement so I wouldn't start my next session with the pillow pattern imprinted on my face.

After dinner, I would often lie on the couch in our living room, desperate, dispirited, and praying for help. I felt defeated as I climbed into bed every night and dreaded the day to come. I kept feeling that I'd reached the end of myself. And yet something kept me going, drawing me back to life when all I wanted to do was give up and fade away.

I was where so many people I had treated had been when they first came to my office. They were beaten down by depression or strung out by anxiety, and a big part of them was ready to throw in the towel. But a bigger part of them was determined to go on. At the core of their being was a fundamental wholeness that had compelled them to seek help in spite of their hopelessness.

They might have felt nothing but darkness inside, but I could see clearly a light that hadn't been dimmed, as if through a crack in their obvious struggles and pain. No matter how they were feeling, seeing that light always gave me hope and even made me smile inside. I knew their suffering did not have to be the end of the story. And I knew that their path to healing had started well before they walked through my door, because the power to heal doesn't start with finding the right treatment. It comes from a place deep inside us.

One night, I finally recognized in myself what I had seen in so many people I had treated. Everything seemed especially hopeless as I lay on the couch after dinner, feeling as if I were dying. I kept saying in my head, "I've come to the end of myself. I've come to the end of myself." And then, in that moment, I realized: the end of *myself* wasn't the end—it was the beginning of something else, something beyond my physical and mental limitations, beyond illness and depression. With my body feeling broken and my mind in a haze, my spirit was laid bare.

This experience brought me back to the most meaningful dream I had ever had. I had woken up crying. My wife, a light sleeper since our kids were born, stirred beside me. "What's wrong?" she asked.

"I dreamed I died," I replied.

"I'm sorry," she said sleepily, reaching out to pat me.

"No," I said, the scene still fresh in my mind. "It was beautiful."

In the dream, the pilot had badly botched our plane's landing. We were off-kilter as we approached the runway, the left wing higher than the right. One wheel touched down before the others, throwing the plane off-balance and causing us to skid across the runway. The plane started twisting and shearing and then tearing apart from the front to the back, where I was seated in the last row. Seats and luggage and the passengers in front of me were flying through the air. I was terrified, expecting the plane to explode at any second, extinguishing my life.

But before it did, I decided to accept my imminent death. I wanted to open to it if it was unavoidable, rather than dying afraid. Clouds of dust and debris washed over me as I leaned back and closed my eyes. I brought to mind the faces of my children so I could die thinking about what I love. Their images filled my mind and heart as I waited for death like waiting for sleep. I was euphoric, knowing implicitly that I was going to join everything I love.

When death came, I felt no pain and no break in consciousness. The color behind my eyelids seamlessly became the purplish space in the night sky, which I was passing through into the stars. I sensed that the spirits of everyone I loved, dead and alive, were there, and I was joining them.

And then I woke up, beside my wife, our kids asleep down the hall. I was crying not because dying was sad but because it was glorious. Experiencing my greatest fear led to the realization of an eternal connection to all that I cared about. There was no place left for fear. More than anything, it was an experience of deep peace.

Remembering this dream, I realized that I had come to the end of myself, but this end meant the beginning of something new and transcendent, as in the dream. I felt a powerful sense of peace that night on the couch and a healing presence within myself. I'd woken up to the fundamental truth of who I am: a spiritual being connected to the divine. And I knew that divine spirit was what I had seen and felt so many times in my patients. That spirit within had kept calling me back to life, in the same way that my patients' spirits had called them to keep going and had called them to the work of therapy.

I had discovered firsthand the constant call of our spirits—a call to thoughts and actions that lead us to wholeness. "I have nothing left," we say. And our spirit answers, "I know. I see your struggles every day, the ones nobody else knows about. It's okay. Come as you are. Life doesn't have to be so hard."

My religious understanding was shaped by Christianity and secular Buddhism, but when I use the word "spirit," I am not assuming a particular religious meaning. "Spirit" is simply the best term I've found for the inner presence I have encountered with people in therapy and in myself, which guides us toward wholeness. Most of us have deep intuitions about this part of ourselves that is neither mind nor body and that is central to who we are. In a way, it is the "you-iest" part of you because it has always been with you and isn't tied to your changing roles, your passing emotions, or your thoughts or actions.

This revelation on the couch was far from the end of my struggle, and it certainly wasn't the last time I would need to hear that call from within. But it was the beginning of hope. It also marked the beginning of a profound shift in how I thought about therapy. For the past several months I had found the practice of CBT to be limiting and considered abandoning it in favor of an unspecified "deeper" approach. But CBT is a powerful method, and I recognized what a loss it would be to give it up. I couldn't forget the faces of the women and men whose lives had been changed through their efforts in CBT.

And yet I knew that I had to go beyond understanding principles and applying techniques. To realize the full potential of CBT, I would need to integrate my training with deeper spiritual truths.

Co-creating Our Lives

Years before my personal crisis, I sat in my office at Penn, looking out my window at the Philadelphia skyline. A red-tailed hawk flew into view, circling higher and higher over the city with only an occasional flap of its wings. I stopped writing whatever grant or paper I was working on and watched until it was nearly out of sight, mesmerized by its effortless flight. Later, I learned from my bird-loving wife that the hawk was riding thermals, powerful updrafts of warm air.

Many birds use thermals to save energy, especially during long migrations. The broad-winged hawk relies on them to travel over four thousand miles as it migrates from the United States and Canada to Mexico and Central America, averaging about seventy miles per day. Without these currents of air, the trip would be quite a slog, taking vastly more time and energy. The hawks would feel every mile. Every day would be a struggle. They might long for rest. And maybe they would despair, in a bird sort of way, of ever making it. Many probably wouldn't survive the journey.

That's what our lives can feel like at times, when everything is hard and every day exhausting. We feel every bump in the road. We're giving all we have, and still it seems it's not enough. We fear for our lives. We're tempted to give up. And then there are those moments when everything stops feeling like such a struggle. We feel buoyed, inspired, uplifted. Life feels more like a dance than a wrestling match. We find flow. That's what our spirits offer us. They are thermals that lift us up when we're overwhelmed and exhausted. Through that spiritual connection, we can find grace and ease.

Hawks and eagles and other birds don't just fall out of their nests into thermals or happen upon them by chance, because the stakes are too high. Birds actively seek them out as collaborators with the air currents. Scientists aren't sure how birds locate thermals, but we know birds are highly attuned to them, as if their lives depend on them. Once they find a thermal, they skillfully navigate the current to stay in it as long as possible.

The same is true for our spiritual connection:

Our spirits provide the will.
Our efforts provide the means.

We need both spirit and effort to live the lives we know are waiting for us. Through our thoughts and actions, we join our spirits in co-creation of our lives. Our spirits can lift us up. We allow ourselves to be lifted. Our spirits continually call to us. We choose how we answer.

The practice of listening for the call of our inner voice, or spirit, is what many call *mindfulness*, and effective therapy is a way of answering that call. Through mindfulness-centered CBT, we can remove the habits that disconnect us from who we are and replace them with thoughts, actions, and mindful awareness that nourish our whole beings, which allows us to stay in touch with that voice of healing within us. The full range of our experience becomes seamless as we align mind, body, and spirit. Healing and ease flow from alignment as we rediscover our wholeness. We stop flap-flap-flapping our way through life and realize we can soar.

Think Act Be

My spirit was drawing me to the work I needed to do to heal from depression. I wanted to feel well again, both for my own sake and for my family's. I missed being involved as a parent and talking with my wife and kids. I was tired of sitting out most of the family activities. I felt bad that my struggles made things harder for my wife. And I wanted to have friends again. All these things meant I had to climb out of the pit of depression.

Well-intentioned friends gently suggested I give antidepressant medication a try, but I knew CBT was what I needed. It was time to take my own medicine. The real strength of CBT isn't in just knowing how it works but in putting it into action day after day.

I was a textbook candidate for the treatment: my mind was filled with self-hating thoughts; I'd given up almost all rewarding activities;

and I was struggling bitterly against the reality I found myself in. I would need to bring my whole being to this work.

- *Think:* my mind needed to change my thoughts.
- *Act:* my body needed to take action.
- *Be:* my spirit needed to find presence and acceptance.

The cause of my physical illness was still unknown, but my mental and emotional healing had already begun. So I made a plan to do self-guided CBT—but a different kind from what I was trained in.

Cognitive and behavioral techniques had formed the core of my work as a therapist. Sometimes I would also introduce mindfulness if it seemed like a good fit for what a patient was dealing with or if the patient expressed an interest in it. But now I realized I was selling mindfulness short by treating it as an add-on. The quality of our presence affects everything. It forms the foundation for all that we think and do. And through openhearted presence, we connect with the deepest parts of ourselves and our experience. In short, we find spiritual connection.

I had felt the imperative of connecting with my spirit ever since that night of despair when I'd reached the end of myself. My spirit had whispered that all was not lost, had shown me that I wasn't hopelessly broken, and had drawn me to the work I needed to do. I wanted nothing more than to stay connected to my spiritual core, and mindful presence offered a way to maintain that connection. Mindfulness would be at the heart of my integrated approach.

As I said, I've come to call this approach Think Act Be. I like to keep things simple, and these three little words capture the power and simplicity of mindfulness-centered CBT. Together, they form the head (*think*), hands (*act*), and heart (*be*) of the practices that helped bring me back to life.

As I used the skills I'd taught so many others, I found the same benefits they had described.

With my head—the *think* component—I monitored my internal voice and realized how fearful and self-critical it was. I found I was constantly telling myself that I was screwing up and was going to lose everything I cared about. What a difference it made to preempt those distorted negative thoughts with life-giving ones that were based in reality.

Through my hands—*act*—I found ways to do more things that brought enjoyment and a feeling of accomplishment. It could be as simple as making our kids' snack, catching up on laundry, or eating my lunch outside instead of at my desk. They were exactly the kinds of things I worked on with my patients to systematically enrich their lives. As unmotivated as I felt at first to make many of these changes, I found my mood lifting—not all at once but gradually and steadily.

And with my heart—*be*—I found ways to bring greater mindful awareness to my day, not only through meditation but in any daily activity. This third element of Think Act Be wasn't an afterthought but the context in which I examined my thoughts and designed my behaviors. Mindful awareness enabled me to see through my distorted thinking and helped me to get the most out of my activities. I was reminded again and again that everything changes when we open to life in this moment, exactly as it is.

I finally had an answer to a question I'd often felt but never fully formed: *Why was I often a bit self-conscious—almost apologetic—about traditional CBT?* I knew it worked, and I'd seen it change lives. But now I realized that I was sensing a gap between the depth of human need and my version of CBT. It was as if I knew implicitly that my understanding of CBT was incomplete. It felt rather thin, like all head and hands, with the heart missing. With the integration of mindful presence into the practice, I felt a passion for CBT that had eluded me. And I wanted to share it with everyone.

What's Ahead

In the pages that follow, I'll present an in-depth view of the Think Act Be model and how to use it. You'll see how these three reminders help us to live in alignment with our mental, physical, and spiritual needs. We know when we're aligned because predictable experiences follow: tension drains from our body; our habitual resistance to reality falls away; we feel unrushed, at peace with ourselves and the world; life is simple; right action flows. There is clarity in knowing and feeling that truth. It's where we experience love, from and for others. We find love for ourselves there, too.

You'll also discover how this approach can inform every part of our lives, from taking care of our bodies to doing meaningful work. At the core of Think Act Be is the recognition that our deepest values can drive the seemingly small decisions we make every day—how we plan our schedules, what we eat, how much we use our phones— countless choices that move us either toward or away from our spiritual core. Creating a world of love and purpose builds from tending to the individual moments of our lives.

The approach I'll be sharing is not about adopting a particular creed or replacing your deeply held convictions. I won't be trying to change your religious beliefs or convert you to my version of spirituality. Rather, Think Act Be helps you to follow where your spirit is already leading, and it can be integrated into any religious tradition.

Through these practices, we can live calmer, more centered lives. We can counter the continual pull of distractions that disconnect us from ourselves, others, and our world. We can give and receive love. And we can find an unconditional peace that's available not just in mystical dreams of death but in every waking moment, even when it seems as if we're broken and will never feel whole. Think Act Be is an invitation to the fullest expression of who we are.

$$\diamondsuit \; 2 \; \diamondsuit$$

CONNECT WITH YOURSELF

After I had been sick for a few years, I began to feel like a different person from who I thought myself to be. In fact, I came to think of this new version of myself as "the impostor." This impostor had taken over my body and mind. "Who is this guy?" I would often wonder. I had always enjoyed making people laugh and spending time with friends, but the impostor was plagued with a chronic seriousness and avoided social contact as much as possible. The real me had had plenty of energy to play with his kids, but the impostor could barely get off the couch. I no longer recognized myself in the person I'd become.

I didn't realize at the time the profound cost of losing connection with myself, though it is easy to see now as I look back. The first step of Think Act Be is to connect with yourself, which is foundational for everything else. Getting out of touch with yourself is a major problem, as I soon discovered.

As time went by, it got harder to tell my real self from the impostor. I started to identify with a diminished version of myself who was defined by struggle and fatigue. One day, my friend Zach asked me how I was doing as we sat together on the edge of our local swimming pool. As I listened to myself describing my ongoing health challenges, I suddenly wondered if the impostor was the real me.

I interrupted myself and turned to face my friend. "I wasn't always like this, was I?" I asked him. Zach had known me when my health was better, and I was hoping he could help me remember my old self. I was starting to doubt that person had ever existed. Maybe I'd always been sick and weak and kind of a downer. I was desperate to rediscover the self I'd always known.

Zach reassured me that I hadn't always been sick, but still I felt alienated from myself. It wasn't just that my illness and depression had changed me. I'd been through other major transitions in my life, such as becoming a parent, and had kept my sense of identity. This was different. It felt as if I'd lost myself—as if my core identity was not just different but was gone.

I caught other glimpses of my former self in photos and videos and remembered that I used to laugh and dance with my kids. My wife also let me know that my struggles weren't the sum total of my identity. But I needed more than these external reminders, more than the knowledge that there was more to me than my confusion and pain. I needed to reconnect with myself.

The most powerful reminder of who I am came from within, at the water's edge in Cape May, New Jersey, where I was vacationing with my family. I had no energy at that point and struggled each day to make the short walk to the beach and back with my family. I had to stop and rest after climbing the few stairs that took us over the sand dunes and onto the beach; once we got set up, I mostly just lay in the shade of our beach umbrella, feeling despondent and confused.

One afternoon midway through the week, I managed to make it into the water, at my wife's urging. "It'll be good for you," she said. "You're never happier than when you're in the water." I grumbled silently to myself as I grabbed my goggles, left the semi-comfort of my towel, and waded into the cool water of the Delaware Bay.

Not long before our Cape May vacation, I had watched the Disney movie *Moana* with my family. The title character is the daughter of the village chief in the islands of Polynesia; she feels called to the sea, but she is forbidden from entering it because of terrible things that happened in the water long ago. And yet the call persists, coming from a part of herself that won't be ignored and inviting her to reconnect with her true identity. In the powerful climactic scene, haunting music plays as Moana sings, "This is not who you are. You know who you are."

As I looked across the expanse of water to the horizon, words from that scene in *Moana* filled my mind. I know it's a Disney movie and I'm a full-grown man, but the words hit me hard and I started to cry as I stood there on the sand. *This is not who you are.* My spirit was speaking to me again, calling to me as it had that night on the couch a few weeks before. *You know who you are, Seth. And this is not it.*

I wasn't the weak and pathetic man I saw myself to be. I wasn't a net negative for the world and my family. I was Moana, daughter of the village chief, descended from voyagers. Just kidding—I was myself, the person I'd been my whole life. I wasn't forgotten or defeated. Yes, I was dealing with an illness. Yes, I was exhausted. This version of myself was having a rough time. And yet I had rediscovered the truest part of myself, which wasn't defined by the difficulties I was experiencing. On a fundamental level, I was whole.

I waded farther out, every step a renewal. When the water was at my chest, I dived under and began to swim. I felt like a new creation as my body slipped through the waves.

There was still a lot of work ahead. The quiet voice that called me back to myself wasn't just helping me to rediscover my true identity. It was calling me to action—to live in alignment with who I was. As I would soon discover, my alienation from myself had started long before I got sick.

Tune In to Yourself

I am certainly not alone in losing myself during a difficult period. Anything that turns our life upside down can lead to a loss of connection with ourselves: losing a relationship through death or divorce; an all-consuming job; addiction to alcohol or drugs; a psychiatric illness. There may even have been no obvious triggering situation or event—we simply drifted away from ourselves by degrees without realizing it. Maybe we can't ever recall really knowing ourselves.

My patient Sarah lost herself somewhere between her mother's untimely death and the trauma of her subsequent abusive relationship. When I first met Sarah, I was struck by how well her life was put together, in spite of the grief and terror she had been through as a young adult. But in the sessions that followed, it became clear that she had very little sense of self. Sarah had learned that her own needs were unlikely to be met. She found it safer to focus on the expectations of those around her rather than ask others to respect her needs and risk being disappointed yet again.

One of the mindfulness practices I introduced that Sarah found most helpful for connecting with herself was the deceptively simple "I Am Here" meditation.[1] It can be done anywhere and in any body position—standing, lying, sitting, walking—with eyes open or closed. As Sarah gently inhaled, she would silently say, "I am." As she breathed out slowly, she would say to herself, "Here." Those three words paired with the breath were a powerful antidote to Sarah's alienation from herself and her experience. The "I" brought her back to herself, "am" tied her to the present, and "here" located her exactly where she was. No matter what she was doing, she could quickly re-establish connection with herself through a single breath cycle with the "I Am Here" declaration.

Find a comfortable place to sit, and set a timer for three minutes. Pair each breath with "I am here" for the duration of your session. You can also do this practice on the go when you're rushing from one thing to the next and it's easy to lose connection with yourself.

As I led Sarah through training in mindful presence, she learned to stay with difficult emotions that she had habitually tried to ignore. She also confronted the traumatic memories of abuse at the hands of her ex-husband. It was difficult work as she peeled back the layers of pain that had accumulated over the years.

In the process, Sarah rediscovered herself. She remembered the strength she carried inside her—the same strength that had supported her through the sudden loss of her mom and through years of abuse. "Strength" is actually putting it mildly; Sarah remembered that she was *fierce*. Only a fierce woman could have left a manipulative and abusive husband who constantly told her she was "no one without him." She had nowhere to go when she left him, and yet she knew that anywhere was better than staying in the hell of her marriage and suffering more beatings and belittlement.

My work with Sarah didn't end when she reconnected with herself. Just as it was for me with my baptism in the Delaware Bay, finding herself again was a crucial step of a larger process for Sarah. In the second part of our journey together, Sarah realized that she needed to make some real changes in her life. She could no longer ignore her needs or let others make choices for her. She would need to act on the inner strength she had reclaimed.

It was a remarkable thing to see this young woman fully inhabit her personhood. Soon Sarah began to surprise the people around her with her directness. She stopped saying yes when she meant no, such as when her partner and daughter once again left a big cleanup for her after a nice dinner that Sarah made. Rather than doing all the cleaning herself while silently resenting her family members, she told them exactly what she needed them to do: clear the table, empty the dishwasher, and put away the leftovers. I was continually amazed by the transformation I was seeing. Was this the same person who had always deferred to others just a few months before? Sarah's new behaviors reinforced the connection she had made with herself and repeatedly served as a striking reminder of who she was.

Think back to a situation in which you said yes when you wanted to say no to someone, such as a demanding boss, an overbearing family member, or a challenging friend. Consider what made it hard to say no. What do you think might have hap-

pened if you had? What did it feel like to go along with something you didn't want to do?

As Sarah and I both discovered, finding ourselves is not simply an intellectual experience of knowing what we're like. It's a living relationship. And as in any strong relationship, it begins with listening. There is a call within each of us that's always present, like a tuning fork that constantly rings true. Finding ourselves always begins with listening for that pure, clear note. As we tune in, we discover what our needs are and what actions we need to take.

This type of listening depends on what I call our "presence," because we can hear only what's happening inside ourselves right now. Therefore, the first step in Think Act Be is simply connecting with ourselves in the present. This quality of connection helps us to know ourselves and our needs and to choose our thoughts and actions accordingly. You'll learn this process well because it recurs over and over throughout this book: *ground yourself in mindful presence, and then think and act from that place.*

This approach is about much more than fixing symptoms and curing illness. Like any system for living well, mindful cognitive behavioral therapy (CBT) works best when we apply it to all aspects of our experience rather than cordoning off "therapy time" from the rest of life.

Transcend the Medical Model

Clinical psychology—and CBT in particular—is steeped in the medical model, with its focus on assessing symptoms, determining a diagnosis, and choosing the right treatment. I learned CBT as a set of simple practices for identifying and alleviating symptoms, which it certainly is. The emphasis is on finding one-to-one fixes that match treatment techniques to a person's condition: Depressed? You need behavioral activation. Anxious and avoidant? You need exposure therapy.

The template for the CBT I learned was medication for psychiatric conditions. Research trials often pitted equivalent "doses" of a psychotherapy and a drug over several weeks of treatment. Our goal as researchers and clinicians was fewer distressing symptoms: less depression, less anxiety, less sleep disturbance. Zero was the best score a participant could receive on any measure.

The medical model of therapy assumes that CBT has little to do with our deepest and most meaningful experiences. Like a pill, it is taken when we're unwell and put away when we're feeling better. We don't rely on CBT for deeper personal growth, any more than we expect aspirin to elevate our daily lives beyond simple pain reduction.

But the medical model is also limiting. In this model, each of us is like a car. Our car's baseline is set when it comes off the assembly line, and maintenance aims to get our car as close to that baseline as possible. The best we can hope for is that our car doesn't break. Thus, CBT for depression is successful if we feel less depressed; CBT for anxiety is successful if we feel less afraid.

But the full scope of the living, breathing, dynamic human experience can't be captured by a mechanical metaphor. We are more like gardens than cars. We grow. We adapt. We need tending. And like gardens, our lives can be productive. When we're healthy, we yield a good harvest, and the fruit of our actions can make the world a better place for others.

Like gardening, taking care of ourselves is a collaborative partnership. Gardens rely on us for their creation, but forces beyond our power and understanding are at work as seeds germinate and leaves photosynthesize. In the same way, we have a life force of our own. We eat nourishing foods and our bodies know how to use them for all of our physiological needs. We face our fears and they diminish. We quiet our bodies and minds and we find sleep. We learn new ways of thinking and our depression lifts. Life-giving change requires work as we offer our minds, bodies, and spirits the right conditions to meet their needs. But there's an ease in our efforts because we're not laboring alone.

As I engaged in mindfulness-centered CBT, I felt firsthand its full potential, which was not simply to alleviate depression or dampen anxiety. Those were important ends but were in the service of something beyond simply not feeling bad. I found that the nuts-and-bolts techniques of Think Act Be were perfectly relevant to the most meaningful parts of my existence, just as I had witnessed many times in my therapy office.

We don't have to put a low ceiling on CBT, deciding that it ends at some arbitrarily defined level of "normal." The benefits of being present have no boundaries. Thinking well is valuable for all of our experience. Acting with purpose is always useful.

In fact, Think Act Be is often most useful when there isn't any discernible problem or diagnosis. For many of us, the problem seems to be precisely that there is no identifiable problem but we can't say that all is well.

My friend Jen described reaching this place as a young adult. All seemed to be going well in her life—she'd finished graduate school, gotten married, and started a family. She had a good job and a nice house. But one day, while her husband's family was visiting, she felt an emptiness that she couldn't account for. As she sat with the other adults watching the kids play on the floor, she was gripped by the question "Is this it?" Somehow, she had expected there would be something . . . more.

Jen's question captures one of the most common and vexing problems we face. We know there's more to life than we're experiencing: More joy. More peace. More love. We want to live more fully and find deeper connection to other people and our experiences, but we're not sure what we need to change. And no matter how much we fill our lives with things and relationships, often our quest for more leads to less.

Thankfully, Think Act Be isn't just for treating disorders—it's a wise system for living well. At its best, it can lead to deep personal growth in all areas of our lives, from how we treat our bodies to our most important relationships. The starting point is wherever you find yourself, and there's no ceiling. Mindfulness-centered CBT is a spiritual practice, if we allow it to be, and a powerful way of finding peace.

I still believe that CBT is a wonderful treatment for circumscribed problems such as anxiety and depression, and I'll keep using it in those contexts. One of the best parts of the treatment is that it can reduce symptoms quickly. But the usefulness of CBT doesn't end when our problems resolve. We can aspire to something beyond the absence of illness—far more than simply being symptom free.

The challenge, then, is to understand who we are and what we need.

Align with Your Truth

When I first heard the call of my inner self, I thought it was beckoning me to become "more spiritual," as if my ideal self were a disembodied spirit, floating through life unfettered by thoughts and untouched by feelings. But idolizing our spiritual selves would mean abandoning our minds and bodies and becoming *less* of who we are. In truth, our spirits are not narrowly focused on "spiritual" things, to the exclusion of our physical and mental realities. As the scientist and theologian Ilia Delio wrote in *The Hours of the Universe*, "the divine dimension of reality is not an object of human knowledge; it is, rather, the depth-dimension to everything that exists."[2] Our spirit permeates all of our experience and is intimately connected to the rest of us.

The practices of Think Act Be help us to align with what is true about us—with that clear note that rings inside of us. "True" here simply means aligned with the reality of who we are in all of our dimensions—mind, body, and spirit. These three dimensions—head, hands, and heart—work together to make us fully alive human beings.

Knowing what is true about ourselves does not need to be deep and mysterious. My wife reminded me of a truth about myself when she encouraged me to go swimming at Cape May. She knew I am happiest when I am swimming; that is one of my truths.

Discover some of your own truths by considering times in your life when you have felt the happiest or most contented. How might you create more opportunities to do the things that bring you real joy?

We know mental truth when we practice right ways of thinking that bring us happiness and peace. We can find that truth in our relationship with our thoughts as we see through the endless fictions our minds create that can make us miserable, such as that we don't deserve to be happy and that we aren't lovable. The cognitive part of CBT helps us to replace these false beliefs with ones that are faithful to reality. We find joy as our mind dwells in the truth.

We experience physical truth when we give our bodies what they need and consistently do the things that bring us alive. We can enact truth through eating nourishing foods, getting adequate rest, moving our bodies every day, spending time with our favorite people, being of service, and doing work we enjoy. Importantly, Think Act Be goes beyond the simple awareness that certain actions are better for us than others. Advice alone does little to change our behavior, as you probably well know. Instead, as we'll see, it offers practical tools based on the science of human behavior to help us make lasting changes that create a more rewarding life.

We find spiritual truth through being wholly present in our lives because our spirits are always in the here and now of our experience.

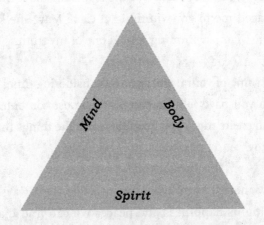

Figure 1

My swim in the bay was an encounter with spiritual truth as I reconnected with myself and what I love. As my example shows, mindful presence is not an esoteric experience that's available only to a select few; being in our lives is a habit that all of us can cultivate in each moment.

Our mind, body, and spirit form an integrated whole, intersecting with and affecting one another (see figure 1). For example, our bodies affect our minds, as when we're well rested and it's easier to recognize our negative thinking. Our bodies affect our spirits, too, as when we step out of compulsive activity and thereby enter into connection with our spirit. And our spirits affect our minds, as when we focus our awareness on the present and discover that in doing so, it's easier to recognize when our thoughts are telling us lies.

The fruits of our collaboration with ourselves often include such benefits as a healthier body, peace of mind, sound sleep, and loving relationships. But answering the call within isn't necessarily about getting back to our former health or abilities. No matter what struggles we may be having, we can offer ourselves what we need right where we are.

Discover Your Care Instructions

The packets of seeds I buy for my garden usually describe the plants' ideal growing conditions—how and when to plant them, whether they need full sun or partial shade, the best kind of soil and fertilizer. These instructions help me to provide my plants with what they need so they will grow up strong and produce a good harvest. The integrative psychiatrist Dr. Omid Naim recommends that we identify our own "ideal growing conditions," just as we do for our plants.[3] Each of us can ask ourselves, "What are the conditions that help me thrive—the soil, sun, and water of my well-being?"

This metaphor resonates not only with my experience as a gardener but also with what I found in my own life. In hindsight, the way I was living before my illness looks like it was *designed* to create poor health. I was neglecting my sleep, my diet wasn't great, I was drinking too much alcohol, and I worked too much with too little time for rest. If I had been a plant in my garden, I would have been living in a shaded patch of earth that rarely got watered. I was missing many of the basic things we need to feel whole.

The same has been true for most of the people I've treated. They often weren't just struggling with unhelpful thoughts or overwhelming fears. Major pieces were missing from their lives, such as meaningful relationships or adequate rest. Narrowly focused CBT techniques, such as questioning their assumptions, fell short of the life changes they needed to make. Tending to their fundamental needs—building strong relationships, getting good sleep—formed the soil for their health and well-being.

Mark was a great example of this principle. He came to me with high anxiety and trouble sleeping and wanted to learn mindfulness for stress reduction. I introduced him to the basics of mindfulness and meditation, and he dutifully practiced the things we worked on each week. However, he hadn't found much relief after a few sessions, and I gently nudged him to consider more fundamental life changes he might need to make. As we examined the structure of his life, Mark realized that his work habits were incompatible with his well-being. Seventy-hour workweeks left him virtually no time to be with his family, caused frequent tension with his partner, and led him to drink too much, which in turn hurt his sleep.

We couldn't expect meditation to offset overworking any more than meditating in the desert would relieve thirst. At some point, Mark had to get out of the desert. Not surprisingly, the changes he needed to make weren't easy. His long hours were fueled by a persistent fear of failure, and at first, cutting back his hours felt like giving up. But these changes gave us the opportunity to address the deeper dynamics that were driving his thoughts and actions. As he dedicated more time to rest and to being with his kids, his anxiety and sleep problems greatly improved. Over time, Mark redefined his definition of success, which he recentered on fostering strong relationships with the people he cared about.

Survey your own "lifescape." Where are the sun-scorched or waterlogged areas? It might be soul-sucking work, for example, or an anemic social life. Consider how you would like these parts of your life to be different. Keep these areas in mind while you're reading this book, and look for ways to tend to them with Think Act Be principles.

Like Mark and many others, I needed to make real changes in the way I approached life. Looking back now, I'm not surprised that I wound up sick and depressed. I was ignoring my fundamental needs—for sleep, exercise, nutrition, socializing—and demanding more of my body than it could sustain. I strongly suspect that my physical illness was due in part to my self-neglect, which began long before my first symptoms appeared.

The irony wasn't lost on me—a psychologist battling many of the same issues that he's helping others to overcome. For years, my struggles were a source of secret shame. Why hadn't I realized I had lost my way and changed course sooner? In spite of my professional training, I had a blind spot for my own needs—or, rather, I had never looked at myself or my situation long enough to see the obvious problem areas. It's hard to spot our unfulfilled needs when we're never really present with ourselves.

When I began the work of aligning with my true nature, I used the same approach as when I saw a patient for the first time: I took some time to get to know myself and took an inventory of the major areas of my life. How was my relationship with my wife? Did I have a good circle of friends? Was I getting consistent movement? As I examined my life, I found that in many ways I was living out of alignment with my true nature.

It felt a little funny to be using the same approach through which I had guided so many people—even relying on some of my own books. Nevertheless, CBT works just as well for therapists as for anyone else. And I am endlessly grateful for the renewal I experienced through my self-guided CBT.

* * *

It's all too easy, as I discovered, to become alienated from ourselves and to lose touch with what we truly need. Healing begins when we simply return to ourselves and tune in to what our mind, body, and spirit are asking for. Through the foundation of mindful awareness we reestablish that internal connection, which reveals the areas of our life that need the water and sun of our attention.

The practices of Think Act Be can be helpful in every area of your life, as you'll see in the chapters ahead. And in contrast with the medical model, the benefits can extend far beyond escape from pain. By offering yourself the right conditions, you can live a healthy and integrated life that is filled with a deep sense of purpose.

With these ideas in mind, let's turn to the fundamental practices from cognitive, behavioral, and mindfulness traditions, which you'll learn about in the next chapter. As you'll discover, these three components address your whole being—mind, body, and spirit—and when joined together offer a powerful way to help you build the life you want.

FIND LEVERAGE

In the previous chapter, we covered how crucial it is to connect with what is really going on inside ourselves if we hope to find wholeness and peace. This chapter provides an overview of how Think Act Be will help us get connected with our experience, as well as specific ways to practice it. In subsequent chapters, we'll drill down more deeply on specific issues and techniques that help us make this healthier way of being in the world a part of our daily lives.

* * *

One afternoon in early summer, I found myself overcome with emotion in the middle of the garden I had built that year. I had dropped to my knees to pull a weed, and the posture made me feel as if I were praying in the warm sunlight. The next thing I knew, I *was* offering a prayer of gratitude. Months earlier, in the cold of winter, I had felt as if I were dying, and now in my garden I felt more alive than I had in years. I had thought I was bringing a garden to life, but I hadn't realized how much the garden was tending to *me*.

This experience represented the consummation of work that began with simple connection to myself and what I cared about. One of the first steps in my self-guided cognitive behavioral therapy (CBT) was to start doing more activities that I enjoyed, since I'd stopped engaging in most of my hobbies. I knew I liked to garden, so I decided to expand my backyard garden and fill it with eight raised beds.

I don't know where I found the energy and motivation to build the beds and fill them with about ten tons of earth, since I was so depleted at that point. I made trip after trip to the hardware store to buy lumber in the early morning twilight. At night, I read books and watched videos about gardening. I grew hundreds of seedlings under the glow of grow lights in the late winter. I had struggled to take care of small chores around the house, and just making dinner at night was often overwhelming. And yet I felt like a man possessed as I built my garden, compelled by something larger than my own efforts.

How was I able to throw myself into the mental and physical work of building my garden when I was so run down? What fueled the labor that brought me back to life and that culminated in that moment of prayer in the sanctuary of my garden? The answer to these questions is the story of Think Act Be, which emphasizes efficiency over raw effort. The work I did on my garden was hard, and yet there was an ease to it. The key was the leverage I found through the tools of CBT.

Harness Mind, Body, and Spirit

The most common challenge we face when we're trying to improve our lives is not knowing what to do but *following through on what we already know*. For example, we may know we need to be more present, improve our diet, or manage our stress, but we struggle to turn these desires into reality.

When we're struggling to follow through on changes we want to make, we often try to muscle our way through with more willpower. We tell ourselves we "just have to try harder" or "be more disciplined." But as you've probably found, willpower is unreliable. Sometimes our motivation is high and sometimes it's low, and it often fails us when we're tired, anxious, or stressed out.

Willpower relies on brute force to help us act in line with our goals; it's like keeping cookies in our cabinet and telling ourselves, "Don't you *dare* eat those cookies." Mounting that ongoing resistance takes a lot of effort, and eventually we're probably going to eat the cookies. No matter how strong our initial motivation, our limited supply of willpower can become depleted, like a muscle that fatigues with use.[1] In contrast, the familiar advice to not keep foods in your house that you're trying not to eat is a simple behavioral tool that provides *leverage*. It requires a single decision in the grocery store to resist cookie temptation, rather than having to say no every time you open your cabinet.

The practices of Think Act Be are first and foremost about finding greater leverage. They amplify your efforts, just as the right tools make your work a lot easier. You *could* clear snow without a shovel or gather leaves without a rake, but with a lot more effort and frustration. With more leverage, you can let go of the struggle.

The same is true of our efforts toward finding peace and living more fully. Cognitive practices shift our thoughts so we're not fighting against our own minds. For example, we can challenge and replace the paralyzing negative automatic thought "Nothing is going to get better, so I shouldn't even try."

Behavioral practices provide leverage through actions that don't require us to fight with our motivation, which makes our efforts easier and more sustainable. For example, I found leverage by choosing to do something I loved—gardening—rather than something I thought I *should* do but that had little appeal.

Mindful acceptance allows us to release our fight against reality and instead to work with what is. When I stopped insisting that I shouldn't be sick, I was able to work more effectively with myself just as I was.

My recovery plan included more than a return to gardening, but this example captures the essence of the path that saved my life. In this chapter, you'll learn the principles and practices of Think Act Be and how you can find leverage for the things you want to do. These themes will recur in subsequent chapters as we explore the model more fully.

Set Your Sights

One of the best ways to find leverage is to set well-defined goals. My rebirth in the backyard began with the goal of expanding my garden. Mindful listening is essential for choosing effective goals. What are our minds, bodies, and spirits asking of us? The goals we set will be more meaningful when they arise from a deeper connection with ourselves. When I turned inward, I felt moved to build a garden.

Good goals inspire us by transforming our vague hopes into a concrete plan and offering a vision of what life can look like through our focused efforts. The most useful goals go beyond "manage anxiety" or "be less depressed" and capture specific ways that our lives will be better. What will we be able to do that we're not doing now? What changes will the people we know see in us? Goals provide a compelling *why* that fuels our actions, as I found repeatedly while building my raised beds and tending to my garden.

Whatever our goals are, in CBT we write them down. I listed all the vegetables I planned to grow and sketched out the design for my raised beds. Writing down my goals made them more real and provided a dose of accountability—reinforcing that I really meant to follow through. I could also return to my list of goals for the garden, checking my progress and reminding myself of what I was working toward.

Consider your own goals for the areas of life that you identified on page 31 in chapter 2. How would you like things to change in each of these domains? How will you know when you've reached your goals? Consider writing them down so you can remember what they are and return to them as often as you want.

Engage Your Whole Being

One of the best sources of leverage in Think Act Be comes from the natural fit of mindful CBT with our triune nature. *Think* engages the mind. *Act* guides what we do with the body. *Be* fosters connection with our spirit (see figure 2).

Cognitive, behavioral, and mindfulness-based therapies were developed separately over the past few decades, beginning in the

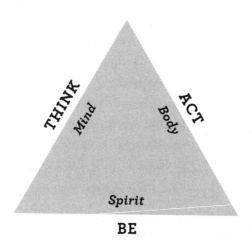

Figure 2

1950s with behavioral techniques. Cognitive therapy came a little later, starting in the 1960s, and mindfulness approaches formed the "third wave" of CBT that emerged toward the end of the twentieth century.

However, their eventual integration was nearly inevitable. Thinking, acting, and being form an intricately connected whole, like rope made of three cords. This holistic approach harnesses all of our resources in the service of a common goal. For example, changing our thoughts often changes our behavior, such as when I stop thinking I'm weak and inadequate and realize I can take action that leads toward my goals. In a reciprocal way, seeing ourselves acting in new ways will change how we think about ourselves, as when I saw more strength in myself as I constructed my garden beds.

Conversely, we hamstring our own efforts when we ignore any of these three components. Thought without action is inert. Action without thought is disorganized. Thought or action without presence is mindless. But when applied together, these three approaches are mutually reinforcing and offer the right tools for our work.

Let's take a closer look at each of the three components of Think Act Be and how they work together to offer us leverage. We'll begin with *being*, the heart of mindful CBT.

Be Present

Living in alignment with who we are begins with mindful presence. "Mindfulness" has become a bit of a buzzword in recent years, with countless magazine covers of blissful people meditating cross-legged on a tropical beach or beside an infinity pool. These clichéd images can obscure the real power of mindfulness: to be fully present in our workaday lives, mundane and messy as they are.

Two simple shifts allow us to step into mindful awareness. We bring our attention to the present, and we open ourselves to exactly what we find there. We'll dive deeper into mindfulness in chapters 4 and 5; for now, let's consider the basics.

ABIDE IN THE HERE AND NOW

I was walking home from the train station one evening when my mind was filled with a sad scenario about one of my kids getting terribly sick. Suddenly I was living in that make-believe world as if it were reality and ignoring the blue sky, the birds singing, and the lovely evening light. I felt sick with worry and sadness, as if the fantasy in my mind were already happening.

It's amazingly useful to be able to imagine things that aren't happening right now. We can anticipate and plan for the future, learn from the past, savor our memories. But this ability has a major downside, as I experienced on my walk home: we can spend most of our lives in a half-dreamlike state. We're captured by thoughts about the future—worrying about what might happen, planning ahead, looking forward to the next thing. Or we're ruminating: dwelling on the past, replaying conversations, feeling guilty about mistakes we've made. And we're living in these dreams as if they're real life, not even knowing we're asleep.

Being constantly lost in the past or the future is like staring at our phones all the time; there's a lot of activity to capture our attention, and yet the scene never really changes. We're at the mercy of whatever our minds imagine, which is often neither useful nor enjoyable. We're worrying about situations that will never happen or regretting things we can't change. Meanwhile, we're missing our actual lives.

In contrast, as the Stoic philosopher Marcus Aurelius said, "If you've seen the present then you've seen everything."[2] That's where life is happening. Our bodies exist only in the present. Real people (not the ones our minds create, like the suffering child I had imagined) are always right here and now. We can experience the world of our senses only in the moment in which we find ourselves. Spiritual connection happens in the present, too.

The most important reason to foster presence is that it allows us to respond to the needs of whatever is before us: "We live in the now," said the Benedictine monk David Steindl-Rast, "by attuning ourselves to the calls of each moment, listening and responding to what each hour, each situation, brings."[3]

The first principle of mindfulness, therefore, is to come into the present. That's where life is happening. That's where we align with our truth—the fundamental truth that we are only ever right here and right now. It takes committed effort to enter into this state of being because our minds continually pull us out of the moment. But as we practice intentionally, we'll start to find ourselves engaging in mindful presence more spontaneously.

There's nothing *bad* about having our attention elsewhere, so we don't have to criticize ourselves for lack of presence; it just hinders our ability to experience as much of life or be available to what each moment is asking of us. And, crucially for Think Act Be, the present is where we can work most effectively with our thoughts and actions.

RELEASE RESISTANCE TO REALITY

The other major facet of mindfulness is letting go of judgments—and as with abiding in the present, this is much easier said than done. Evaluating our situation as *for me* or *against me* tends to be our default mode: "Do I like or not like this person?" "Will this be a good day or a bad day?"

As you pay attention to your mind, you'll find that it's making these automatic judgments all the time. When I was cycling, I found I was carving up my rides into good parts and bad parts. The descents were *good*: the speed, the ease, the cool breeze on hot days. The climbs were *bad*: burning thighs, no breeze, gasping for air. I felt a gut-level judgment that the uphill stretches *shouldn't be there*.

Finally, I realized I was wishing away most of my time on the bike, since going up hills takes so much longer than going down. When I released my internal resistance to climbing, the difference was stunning. I was no longer fighting on two fronts—the physical challenge plus my mental resistance to it—and could focus my efforts on the climbs. They were still grueling, but they were a lot easier when I stopped silently insisting that *this shouldn't be happening*. Even though my external situation looked exactly the same, internal acceptance completely changed my experience of it.

Take a moment right now to notice if you're rejecting reality. Is there something you're actively pushing away or telling yourself "shouldn't be happening"? What are the "hills" in your own life that you find yourself habitually rejecting? Consider what opening yourself to those experiences would be like—letting go of the struggle against your circumstances and instead starting to work with them.

Acceptance and presence are two sides of the same coin. We're fully in the present only when we accept what we find there. As we practice acceptance and let go of resistance to what is, we foster a deeper connection to our experience. Consequently, we're more available to meet the demands of each moment.

BRING MINDFULNESS TO THOUGHTS AND ACTIONS

Mindfulness is the foundation for cognitive and behavioral techniques. It's easier to observe the workings of the mind when our attention is in the present. We can notice the stories our minds are telling us without assuming they're true. When I woke up to the present on my walk home, the sad story my mind had created dissolved like mist, and my mind was filled with the beauty of the scene in front of me. As we'll see in chapter 4, mindful presence even allows us to change the fundamental relationship we have with our thoughts and beliefs.

We'll also see how opening to each moment helps us do the things that are important to us, with less resistance and a greater sense of purpose. The quality of our attention changes everything, even a simple chore like doing laundry. I can put a load of clothes in the washer while worrying about how long it's taking and everything else I have to do, or I can actually notice what I'm doing without resenting the time it takes. Resistance makes laundry a mindless drag, whereas presence and acceptance can make it a strangely joyful experience.

Think Clearly

It's 3:30 p.m.

That car is blue.

I'm a loser.

Obviously, that last statement (which can also come as a wordless, gut-level impression) is different from the first two, which are objectively verifiable facts. There's very little interpretation involved. But thinking "I'm a loser" involves a massive amount of interpretation. We often don't realize when our mind has shifted from reading us front-page news to reading the op-eds. If we don't recognize our thoughts for what they are and treat them accordingly, we'll live in a false reality of our mind's creation.

When I was depressed, my thoughts went in a predictable direction; I was constantly criticizing myself and blaming myself for all of my problems. It was my fault I was sick, my fault I was exhausted, my fault we were struggling financially. My mind convinced me I was such a hopeless mess that it would be better for my loved ones if I just ended my life.

Our minds have the power to affect our emotions and behaviors. Even a passing thought that escapes our conscious awareness can color our experience.

- We tell ourselves, "You shouldn't have done that," and we feel guilty or ashamed.
- We believe the thought "They should be more considerate," and we feel angry.
- We think, "This headache means I have a brain tumor," and we feel anxious.

Beliefs affect our actions, too, as when we believe we're sick and so make an appointment to see the doctor. Our thoughts can even contribute to serious conditions such as anxiety disorders and major depression, as countless studies have shown.

We are constantly thinking; even if we decide to stop thinking, our minds will keep doing it anyway. It's what they are good at. If they aren't telling us stories in words, they're crafting made-up scenes or pulling up images from our memory banks. Our minds are actually so caught up in thinking that we don't realize we're thinking. We assume that the nonstop stream in our heads is something real and meaningful, and we mistake thoughts for actual observations of something true. In subtle ways we probably don't notice, mental events in our brains are fashioning our lives.

Unfortunately, this is often not a good thing because our thoughts tend to be biased in negative directions. We see danger where it doesn't exist. We worry needlessly about the future. We assume the worst about others' intentions. We imagine people are thinking bad things about us while they fear that we're thinking the same about them. Our beliefs about ourselves are usually the most distorted as we exaggerate our defects and minimize our strengths. We might even believe self-hating thoughts like mine that can lead a person to suicide.

The developers of cognitive therapy realized the powerful effects our thoughts can have on our emotions and behavior.[4] They were heavily influenced by Stoic philosophers such as Epictetus, who famously said, "It is not events that disturb people, it is their judgments concerning them."[5] The judgments we make often have a consistent theme that colors all of our experience.

CHECK YOUR LENSES

I was in high school when I got my first pair of glasses. I hadn't known I was nearsighted and was blown away when I put them on. I could see individual leaves on trees! I had assumed everyone just saw blobs of green as I did. My new lenses changed my experience of everything I saw.

My patient Jonathan hadn't noticed the mental glasses through which he was seeing the world. He believed he was pathetic, and he saw everything in his life through that lens. His clothes seemed pathetic. His lunch seemed pathetic. He even thought it was pathetic that he felt pathetic. The "lenses" he wore felt so much a part of him that he didn't even know he was wearing them.

Jonathan's mental glasses reflected a core belief about his inadequacy. In CBT, "core beliefs" are firmly held assumptions about ourselves, other people, and the world. Some may be useful and accurate, and others, like Jonathan's, involve faulty assumptions. As Jonathan and I worked together, he started to recognize the stories his mind was telling and the recurrent theme of his core belief.

It wasn't easy for Jonathan to change his core belief, which he'd had for as long as he could remember. There was no solid evidence that he was pathetic, and yet the belief persisted. Finally, he had a breakthrough during one of our sessions together. He was describing a presentation he'd given at work and how the whole thing had been stupid and weak.

I pressed him a bit more this time—what was the evidence for these assumptions? At first, he seemed annoyed by my persistence because he saw his failures as self-evident. "I understand that you see yourself that way," I said, "but I still haven't heard what was actually stupid and weak about your presentation." Finally, in frustration he blurted out, "I just know that's what I'm like!"

He was staring at his hands in the silence that followed, and I could see that Jonathan was no longer looking through his unhelpful tinted glasses but *at* them. "That's just how I see myself," he said as he looked up at me. There were tears in his eyes. "That's just how I see myself," he said again. "Oh my God." He sounded both horrified and relieved—horrified that he'd been self-critical for so long and relieved to realize that he'd been wrong about himself.

In the weeks that followed, Jonathan was amazed to see how dramatically the world had changed now that he'd taken off his glasses. He could hardly believe all the ways his mind had tricked him into hating himself and how powerful his thoughts had been in creating his reality. As he started to believe in his own worth, he experienced the world and other people in a new way. Nothing had changed in the nuts and bolts of his life, but his perspective had changed everything.

CHANGE YOUR MIND

The key to challenging our false beliefs is to notice them. Thinking more effective thoughts starts with awareness as we realize what the mind is up to. If you want to practice noticing unhelpful thoughts, you won't have to wait long. Your mind will almost certainly tell you a story in the next few minutes, if it's not happening right at this moment. It can be quite powerful just to notice that your mind is crafting a story and not reporting unfiltered news.

Once I became aware that my suicidal thoughts were stories my mind was telling me, I was able to take a closer look at them. I asked myself, "Would my family truly be better off without me?" In traditional cognitive therapy, I would have written down my automatic thoughts and worked methodically through the evidence for and against them: "Am I leaving anything out?" "Is my mind exaggerating some things and ignoring others?" "Was my grandfather's suicide good for his family?"

It can be freeing to liberate thoughts from the mind and put them on paper, and a systematic approach can provide a lot of leverage as we write down our observations and check how well our beliefs align with reality. But, like most people, I didn't find it necessary to go through all those steps in rote fashion. I got the most mileage from *simply seeing the story*, the essential first step in the *think* approach.

Once I had the insight that I was having a thought that may or may not be true, I could consider alternatives to how I was seeing things:

- My family would be worse off without me.
- I actually add value to their life.
- What they really need is more of me, not less.

See what happens when you catch your mind making up stories. For example, your throat might feel a bit scratchy and you'll think you must be coming down with strep throat. At first, you might run with that thought and mentally create a fantasy world in which you're already sick. But then you notice the story and ask yourself what an alternative might be—maybe you just need water or you're having allergies. Suddenly that whole make-believe world crumbles, as quickly as your mind constructed it.

The next time you feel a surge of anxiety or another difficult emotion, pause and take a slow breath in and out. Then ask yourself, "What just went through my mind?" Write down whatever you discover. Notice whether your feelings make sense on the basis of your thoughts, such as sadness if your mind says, "I have nothing to offer." Finally, ask yourself the crucial question: Is this thought completely true, or are there other ways of seeing this situation?

THINK MINDFULLY

When I was sick and depressed, my reflexive response was to think that *this shouldn't be happening.* I shouldn't be sick for no apparent reason. I shouldn't be so tired. I shouldn't have to suffer like this. Those reactions reflected a more deeply held conviction, that *life should be free of problems*—or at least shouldn't be so hard.

This core belief is one that most of us share. We know on one level that life is hard and that suffering is unavoidable. And yet each time we run into a problem, it feels like a glitch in the universe's programming. "I can't believe I have to deal with this," we think.

Just as mindful presence affects our thoughts, believing these kinds of stories affects our ability to maintain mindfulness. We can't be truly present and open to our experience when we're attached to the belief that life is doing us wrong.

One evening as I crawled into bed, I recognized the familiar "woe is me" refrain in my mind and the unhelpful thoughts I was having that led me to resist my struggles. In that moment, I saw those thoughts as just stories and considered other ways I might interpret my situation. Maybe problems are a part of being alive. Perhaps I had what I needed to meet each day's challenges. And maybe peace was available even though things were hard.

Our hearts and heads are mutually supportive: *being* supports *thinking*, which in turn leads us back to being. Heads and hearts need a way to interface with the world, of course, which brings us to *act*, the hands of Think Act Be. Mindfulness and right thinking pave the way for action.

Act Intentionally

When Paula came to me for treatment of her dog phobia, she knew her fears didn't make sense, as her partner had reminded her countless times. And yet her anxiety and avoidance persisted. Extended discussions about fears like Paula's are rarely helpful, so we focused instead on helping her face her fear of dogs. Over time, Paula found not only that her anxiety diminished but also that her beliefs about dogs changed. Where the body led, the mind followed.

Just as our beliefs can shape our reality, our actions can rewrite what we think is true, which is especially useful when rational argument isn't enough to change our minds (as Paula found with her dog phobia). Our brains are constantly making inferences based on our actions: who we are, what's true about the world, what's important to us. If we approach something we fear, maybe it's not actually dangerous. If we treat ourselves kindly, we must be someone worth taking care of. If we invest in a cause, we must care about it. Action changes us. Behavior is therapy.

The behavioral component of CBT is based on well-established principles that date back to antiquity, such as the principle that our unrealistic fears shrink when we face them. These principles were confirmed in laboratory research with nonhuman animals in the first half of the twentieth century.[6] The results from studies of dogs and pigeons are remarkably applicable to the science of human behavior, as the developers of behavior therapy discovered.[7] They created powerful treatments for conditions such as anxiety, depression, and post-traumatic stress based on simple changes in our actions.

The principles of behavior therapy extend far beyond treating psychological distress like Paula's and can offer leverage for any change we want to make to improve our lives. Let's review some of the big ideas of behavior therapy, which we'll examine more closely in later chapters.

IDENTIFY COSTS AND REWARDS

When I was deeply depressed, I was avoiding many things: projects around the house, hobbies, social engagements. There was some relief in not having to do things that seemed like a hassle, but in the process my world shrank and I fell deeper into the pit of depression. The long- and short-term payoffs of my actions followed a common pattern of immediate reward (relief) with delayed costs that were less apparent (depression). I was on the wrong side of costs and rewards, which were shaping my behavior in ways that worked against my best interest.

When I finally recognized the effects of my withdrawal, I made my plan to engage again, which included building my garden. The costs of my action plan were immediate (time and effort), whereas the reward of feeling better developed slowly over the weeks that followed.

The actions you choose depend heavily on how you perceive the balance of costs and rewards involved. By understanding patterns of costs and rewards for your behavior, you can design strategies that help you to align with your true goals. You can build healthier habits. You can plan rewarding activities that will pull you out of depression, as I did. You can face the fears that have held you back from living fully.

The key to behavioral leverage is to make what you want to do as low in cost and as high in reward as possible, which is exactly what the *act* in Think Act Be is designed to do.

GO BIG BY GOING SMALL

When Paula and I treated her dog phobia, she didn't start with petting large dogs. She faced her small fears first, like staying on the same side of the street as a neighbor who was walking his dog. These initial exposures boosted her confidence and reduced her fear. Gradually she worked up to bigger challenges, including visiting a dog park and petting some of the off-leash dogs there.

Breaking down daunting tasks into smaller pieces is one of the most powerful tools of behavior therapy. This approach is directly akin to using a ladder, which converts an impossibly large gap—say, from your floor to the ceiling—into a series of small, manageable steps. As you ascend the lower rungs, the higher ones are easily within reach.

In the same way, nothing is too hard in behavior therapy—it's just too big. By going small and reducing a large task into a series of little ones, we can do things we thought were impossible. Once we start, we're very likely to continue.

Think of something you've been putting off, like a daunting challenge or a dreaded chore, and write it on a piece of paper. Next, write down the first small step toward completing it. Be sure to make the step very manageable, even if it seems ridiculously small. Is it possible to complete that first step soon, even today?

It's easy to discount the value of small, easy actions. How could they possibly change anything? But life is a series of small actions, and although a single constructive act may seem like no big deal, never doing it is a huge deal. Getting off the couch one time to work on my garden wouldn't have cured my depression, but continuing to withdraw would have driven me further into despair. Facing your first small fear won't revolutionize your life, but never taking that step will keep you stuck and afraid. A single breath is a trivial act, but not breathing is an emergency. We accomplish the big things in our lives by doing the little things one small step at a time.

BE CONSISTENT

Inherent in the power of small actions is the need for consistency. Saving 5 percent of your income this month won't build a nice nest egg, but putting away 5 percent each month for forty years will. If you summon your courage and approach what you're afraid of one time, the fear will almost certainly stay with you, but confronting your fears again and again will wear them down.

Accordingly, behavior therapy offers leverage through consistency. One of my CBT goals was consistent exercise, so I scheduled specific workouts for specific times and put reminders in my calendar in case I forgot. An accountability partner can also help with consistency, as when my wife and I scheduled times to do yoga together; making commitments to another person can help us keep the promises we make to ourselves. You can also reduce the cost of actions you want to take when you know your motivation will be low; for example, you can set out your clothes and shoes the night before so the barrier to your morning workout will be as low as possible.

ACT MINDFULLY

As I started my own CBT, I realized that excessive work and too much screen time were cutting me off from mindful presence. Adding more meditation to my schedule wasn't the solution; I needed to align my daily habits with my mental, physical, and spiritual needs. That alignment in itself was mindful action. As I simply did what needed to be done, I naturally entered into a state of greater acceptance and awareness.

While in principle anything can be done mindfully, misaligned actions will interfere with mindful presence. When we neglect our sleep, for example, or put off work we need to do, we're resisting reality through our action—the reality of our sleep needs or of our work. As a result, we stop being truly present. Even meditation stops being a mindful activity if we use it to avoid things we need to take care of.

Mindful presence guides right action, which in turn can foster greater presence. In Paula's treatment, for example, mindful acceptance of discomfort helped her face her fears, which led to her being more fully present in her life. We often think that practicing mindfulness has to involve formal activities such as meditation or prayer, but anything we do is an invitation to be present and available. As we open to what's before us, our everyday actions become spiritual practices.

Putting It All Together

I've described the three components of Think Act Be separately, but their real power comes from using them together as an integrated whole. Years ago, I experienced this power in my office. It was the Friday evening of another busy week, and I was feeling overwhelmed as I thought of all the tasks I hadn't completed. I was thinking I would just have to stay late to finish them even though I was exhausted and just wanted a break.

But then I became aware of how frantic I felt, which I knew wasn't a good place from which to act. So I took the first step in Think Act Be: I came back to myself and to the moment. I placed my hands on my desk to connect with something real and present, and took a slow breath in and out. Immediately I felt a bit of tension dissolve and felt more hopeful that my evening might not have to be as punishing as I thought.

Now that I was a bit less stressed and more grounded, I asked myself what story my mind was telling. I realized my thoughts were filled with "have-to's": "I have to finish this blog post." "I have to return these emails." "I have to read this chapter." As I looked at the thoughts more objectively, I realized that none of the tasks were as urgent as I thought they were, and none had to be done that evening. A more helpful alternative that I told myself was "I don't have to keep pushing when I'm already exhausted; there will be time for all these things."

Finally, I used my mindful awareness and new perspective to take appropriate action. That evening, it meant making a schedule for each of the things I had to do so I knew I would finish them on time, and then packing up my things, riding my bike home, and starting the weekend with my family. The whole process from getting grounded to choosing action took about two minutes, and it completely changed my Friday night. Instead of greeting my wife and kids with stress and exhaustion, I was able to meet them with excitement about our weekend together and the peace of knowing my tasks would get done.

Come as You Are

The Roman emperor and Stoic philosopher Marcus Aurelius summarized the essence of Think Act Be nearly two thousand years ago:

> *Objective judgment, now, at this very moment.* [think]
> *Unselfish action, now, at this very moment.* [act]
> *Willing acceptance—now, at this very moment—of all external events.* [be]
> *That's all you need.*[8]

Aurelius even emphasized the centrality of presence: judgment *now*; action *now*; acceptance *now*.

I'm often amazed when I read texts from long ago that precisely describe the principles of modern mindful CBT. But then again, maybe it shouldn't be surprising. Every good idea seems to be a rediscovery of something that's long been known. Like most things that are powerful and transformative, these are really simple approaches that work precisely because they *aren't* novel.

At the same time, mindful CBT is more than a new packaging of ancient wisdom. The power of CBT comes from integrating age-old ideas with more recent scientific revelations about the forces that drive our thoughts and behavior and the power of mindfulness to fundamentally shift the way we meet the world. The therapy approach that emerged offers a systematic framework for putting principles into practice.

Through this framework, we can discover a whole world that we're often unaware of—a world of our thoughts, of patterns of costs and rewards, of the way we interface with our experience from moment to moment. Attending to these three aspects of my experience transformed my life. Mindful awareness (*be*) ushered me into the present, where I connected with the deepest parts of myself. Training my mind (*think*) quieted the storm of self-loathing in my head, making room for the truth: that the universe had called me into being and that I was embraced exactly as I was. Simple shifts in my behavior (*act*)—such as building my garden—brought me joy and fulfillment. These three components were seamless and mutually reinforcing.

So come as you are. Aligning with your nature starts with coming home to exactly where you find yourself. You have everything you need—your mind for judgment, your body for action, your spirit for willing acceptance. Think Act Be is for every part of yourself and every area of your life.

Now that you have a good sense of the major techniques from cognitive, behavioral, and mindfulness approaches and how they complement one another, let's turn to a deeper exploration of what happens when we stop resisting and start saying yes to our experience. No matter what's hindering us from finding peace or no matter what hurts inside, we begin with present-focused awareness.

4

SAY YES

In chapter 3, we explored the three elements of Think Act Be and how they work together to help us build the lives we want. This chapter focuses on mindfulness, the foundation of the approach, which supports and enhances all the cognitive and behavioral work in mindful cognitive behavioral therapy. We'll examine exactly what it means to say yes to our experience and why this simple response can change everything. In chapter 5, we'll cover specific ways to practice mindful awareness as part of Think Act Be.

* * *

"Oh no," I thought as I stopped typing and squeezed my eyes shut. "You've got to be kidding me." The sentiment was directed at my eight-month-old daughter, Ada. I had thought for sure she was tired enough to fall asleep after my third attempt to put her down for the night. I'd even made it downstairs and resumed my grant writing, but now she was crying again.

I said a silent prayer of thanks for the privilege of being a father as I headed back upstairs to settle her once more. Not really. I was far from happy about having to stop my work and attend to her, and I swore under my breath before going back in for round four of the nightly sleep training.

The whirring of the white noise machine greeted me as I opened the nursery door, along with Ada's crying. *Deep breath.* I picked her up for a moment to help her calm down before laying her back in the crib. She immediately rolled onto her tummy and I resumed patting her back, feeling sorry for myself for being stuck in this predicament.

"I can't believe you're not asleep," I thought as I gritted my teeth in the dark. "I need to get back to work. I don't have time for this. This is ridiculous. You should be asleep already. You're so tired. I hate this. Why won't you just stay asleep?" We'd been doing sleep training for several nights, and every time I put Ada down, I hoped it would be easy. It rarely was, and I felt angry and resentful every time she cried and I had to go back in.

But on this night, something shifted. As I stood in the dark patting my daughter's tiny back, it occurred to me that nothing was really wrong in this moment. I wasn't starving. I wasn't in pain. My family was safe. On the face of it, things weren't so bad. I had a few moments with my baby, who I didn't have much time with during the workweek. She was healthy, which I didn't take for granted after a scare with her heart a few months earlier. She would fall asleep eventually.

How can we find joy even when life doesn't go the way we want it to? My experience with Baby Ada's sleep gave me a powerful answer to this question. I still wanted her to fall asleep as soon as possible.

I hoped I wouldn't have to come back into her room again. But I'd moved from resisting what was happening to accepting it.

Do What You're Doing

The only barrier to my contentment as I tried to help my baby sleep was *my rejection of what was happening*. I was fixated on my plan for the evening and fighting anything that stood in the way. But in a matter of moments, I experienced all that Think Act Be has to offer: a different quality of *being* completely changed how I was *thinking* and helped me align my *actions* with my circumstances. Joy followed.

When I accepted that Ada was awake, I immediately saw through the false stories my mind was telling me, especially my insistence that "this shouldn't be happening." Accepting reality in the context of mindfulness doesn't necessarily mean we *like* a situation or that we're resigned to it. For Steven C. Hayes, the creator of acceptance and commitment therapy (ACT)—a powerful form of CBT—acceptance is about embracing all of our experience. "The word 'acceptance' comes from a Latin root that means 'to receive,'" Steve told me.[1] Instead of fighting against what is, we choose to receive it. And, as Steve pointed out, what we receive can become a gift.

When I stopped rejecting reality, those moments with my daughter *did* feel like a gift. All those times when I'd looked this gift horse in the mouth, I'd been suffering more than I had to, and I was probably radiating that energy to Ada. The worst thing about Ada's sleeplessness wasn't Ada's sleeplessness but my insistence that the situation should be different from what it was. As I opened to what was happening, I simply acknowledged that *this is how things are right now.*

Nothing about the situation changed, but I was completely liberated from the false reality my mind had created.

Most of the time, we're unhappy not because of the situation we are in but because we are *not* really in the situation. Our bodies are present, but our minds are focused on getting away and doing what we think we ought to be doing, as was the case with a friend of mine who loved to run. For years he'd been running six days per week, and he was always doing competitive races. But then he got injured and could no longer tolerate the impact of running, so he switched to speed walking—and hated it. He was usually scowling when I saw him out for his morning walks.

One day, as we passed each other, I asked how it was going. He responded, "I hate not running." I heard the grief in his voice—he was mourning the loss of what he loved. I also realized what he was doing every time he went for a walk—*not running*. Of course he hated it. Walking isn't so bad, but there's no joy in not running. Every part of the experience reminded him of what he wanted to be doing as he put on his not-running outfit and laced up his not-running shoes for his morning not-run.

> **The next time you have to do an activity that you see as unpleasant and tend to struggle against—like washing the dishes or taking out the trash—offer it your full attention for as long as you're doing it. When you notice that you're mentally rejecting the experience, see what it's like to welcome it with friendly acceptance. You don't have to try to make yourself like it; simply receive it as what's happening right now.**

When I was focused on what I thought *should* be happening—Ada sleeping, me at my computer—I wasn't actually doing what I was doing. I wasn't rubbing her back; I was *not working*. I was mad not because Ada was awake and needed me but because my attention was on what I couldn't do.

Broaden Your Ideas About Mindfulness

What I discovered as I stood at the crib that night was a demonstration of what is often called "mindfulness." It's been defined in different ways, but the essence is a focus on the present with openhearted acceptance, as we saw in chapter 3. This stance counters our habitual tendencies to focus on the past and future and to struggle against things as they are.

I hesitate to describe this way of being present as mindfulness because labels are limiting. When you read "mindfulness," you probably imagine someone meditating or maybe doing yoga. It might bring to mind stones beautifully balanced in front of the ocean, or a lotus flower in the water, or perhaps gurus from Eastern religions (especially Buddhism). We also tend to believe implicitly that mindfulness involves certain actions—sitting cross-legged, closing your eyes, taking a slow breath, feeling your body.

These concepts of mindfulness can limit our encounter with each moment, just as the concept of "God" can restrict our understanding of true divinity. When we believe that mindfulness is *this*, we also think it's not *that*: mindfulness is meditating but not talking with our spouse. It's tuning in to our breath but not making dinner, practicing tai chi but not driving to and from our lesson. However, openness and presence are available no matter what we're doing, and just as much for the moments we call "ordinary" as for those we cordon off as "sacred."

Common portrayals of mindfulness can also make it seem very

nice and peaceful and contained—and a bit inaccessible. We know our lives look nothing like those mindfulness magazine covers. In contrast, the actual experience of openhearted presence is anything but tame and predictable. It might be relaxing and peaceful at times, but real mindful awareness can also be raw and wild and unbounded.

This distinction between limiting *ideas* about mindfulness and the *experience* of being open and present is crucial. Otherwise, mindful presence can feel complicated and elusive—like something we need to add to whatever we're doing. But mindfulness is more subtraction than addition, and it is truly the simplest thing: we just say *yes* to everything. We notice the urge to shut down and escape, and instead we receive what each moment offers. Even if we do shut down, we can say yes to that awareness, too: "Right now I am struggling to stay in this moment. I am running away. That's what's happening."

In spite of these limitations, I'll use "mindfulness" throughout this book to describe openhearted presence. Just keep in mind that the label doesn't capture the full experience and that mindfulness need not include any of the trappings we've come to associate with it.

Align with Reality

If my kids are reading quietly while I'm trying to write, reality gets my thumbs-up. (Right now, they are. Good job, universe.) If they're wrestling and screaming, reality is *bad*. When I'm in default mode, I tend to treat everything this way—the weather, the temperature, how people treat me, the state of my body—believing implicitly that I won't be okay until reality bends to my will. That was my stance during my daughter's sleep training because I expected it to go according to my plan.

Our typical mode of saying no to our experience places many expectations on life: *Reality should be different. I must be in control. I demand certainty. Things shouldn't take so long. I must be comfortable.* The common thread in all of these beliefs is an evaluation of life based on whether it obeys our wishes. The major downside to all this naysaying is that it gets in the way of our happiness.

By entering a state of presence and acceptance, we align with the true state of the world: *this is life.* We start to see through the beliefs we had mistaken for reality, as in my assumptions that night in the nursery: "Ada should be asleep. I shouldn't have to settle her again." I thought my preferences were the standard and that reality was sinning against me. Maybe if I expressed my angry wishes forcefully enough, the universe would resolve to do better. When I saw more clearly, I knew my wanting the situation to be a certain way didn't mean it had to follow my rules.

Reality doesn't care about our "shoulds," any more than the weather cares about our forecasts. Who said that Ada *should* be asleep? It wasn't as if she were defying the laws of physics. Maybe it made perfect sense that she was still awake, based on factors I wasn't aware of. The only thing real about our insistence that the world conform to our wishes is the misery we'll subject ourselves to in the process. But when we open to what is, we can find equanimity even in very hard times.

Let Mindful Awareness Change Everything

Not long after the eye-opening incident with Ada, I was treating a young man named Josh who struggled with social anxiety. He was a good guy—a recent college grad working in construction while he looked for a job in his field. He was concerned that his anxiety would hold him back professionally, so he came to me for a few weeks of CBT.

In one of our early sessions, I introduced Josh to the basics of mindful presence in social situations and gave him some homework to practice. In our next session, I asked how it had gone. "Well . . . ," he began, seeming unsure of how to put his thoughts into words. "When I focus on what's happening in my conversations, I'm not obsessing about how I'm coming across to the other person, so I've felt a lot less anxious around people. And I realized that when I just pay attention to whatever I'm doing, like washing the dishes, my mind stops replaying interactions where I might have seemed awkward. I'm even falling asleep more easily when I just focus on my breathing. So I feel like mindfulness . . . kind of helps with everything."

He couldn't have been more right. Mindfulness is similar to good sleep in that it can improve every part of our lives—not always as quickly as it did for Josh, but consistent practice yields reliable results. Research shows that mindfulness training helps to alleviate virtually every condition it has been applied to, including depression, anxiety, obsessive-compulsive disorder (OCD), post-traumatic stress disorder (PTSD), attention deficit hyperactivity disorder (ADHD), social anxiety, panic disorder, eating disorders, borderline personality disorder, alcohol addiction, chronic pain, and insomnia.

Practice in mindfulness can do more than fix psychological disorders. It also offers benefits for positive outcomes such as life satisfaction, relationship quality, stress management, creativity, positive emotions, perceived quality of life, attentional focus, and even experiences of awe. Mindfulness practice can change the physical structure of the brain, and its benefits may extend to improved immune function.

How can one simple practice have such wide-ranging effects? It's hard to think of many interventions that can help in so many areas. The things that come to mind aren't so much "interventions" as they are the foundations of life: Air. Water. Food.

There is something similarly elemental about mindful awareness. The quality of our presence affects all of our experience because it's how we interface with the world. If we had gone through life wearing oven mitts on our hands, removing them would improve our dexterity in every activity. We would be amazed at how much easier it is to type, sort mail, press the buttons on the remote control, thread a needle. In a similar way, being in and embracing the moment fundamentally changes our experience.

This is not to say that greater connection to our experience is always comfortable. Oven mitts are clumsy but protective, and sometimes it's excruciating to be fully present—to stay with our sadness or to turn toward another's pain. If we hope to bliss out through mindfulness, we risk turning the practice into a way of avoiding discomfort rather than opening to all of life. Part of the gift of mindful awareness is helping us see that perhaps comfort isn't the supreme goal in life.

It's natural to prefer that good things happen to us, but there is a fuller, richer way to live than constantly seeking pleasure and avoiding pain. Paradoxically, by being willing to feel everything, we experience less suffering. Being mindfully centered in the present is like being physically centered, with our weight evenly distributed front to back and side to side. We're better prepared to respond to whatever comes at us and less likely to be thrown off-balance. In the process, we awaken to a life that's ultimately more meaningful and satisfying.

> **Think of a time when your mind was insisting that an uncertain outcome—like the results of a medical test or a job interview—must turn out the way you were hoping. How did it feel to be attached to a particular outcome as the only acceptable one? What might it have felt like to open a bit more to whatever the outcome would be?**

Decenter from Thoughts and Feelings

When I'm writing a book, my thoughts and feelings about it vary from day to day. Some days I'll feel optimistic that the book will turn out well, and other days I'll be seized by anxiety and thoughts like "I can't do this." In the past I was highly reactive to these fluctuations, feeling triumphant when I thought things were going well and hopeless when confronted with self-doubt. My well-being was driven by my passing mental states.

Thankfully, I've learned to see those ups and downs as uninformative noise. They tend to be driven by things that aren't directly related to my book, such as my mood and energy level or how stressed I'm feeling. They're truly irrelevant unless I give them more importance than they're due. With mindful presence, I've found some distance from my thoughts and emotions and see them as events that are happening: "Now I'm having feelings of confidence." "Now I'm having thoughts of self-doubt." From this vantage point, I'm no longer strongly identified with what I think and feel, becoming instead the *observer* of my experience.

When I was upset about Ada's sleep, I assumed the situation it-self was irritating, and I was unaware of the mental and emotional processes that were generating my irritation. I couldn't examine my reactions, much less change them, because I saw them as inherent to the situation. I'd been in the thick of a battle, fighting against reality without realizing what the fight was about.

Once I discovered that my feelings were driven by my resistance, I was no longer lost in a fog of difficult thoughts and emotions. Gain-ing greater awareness was like withdrawing to a hill where I could observe the battle and understand the forces at play. That new per-spective offered more freedom in how I responded. I realized that I'd been waging war against my own side and that there didn't need to be a fight at all.

This process is known as *decentering* because passing thoughts and feelings are no longer at the center of our identity. It's a subtle shift that can have an impressive effect on our well-being. Taking even a half step back in our awareness gives us greater choice in how we handle challenging moments. We can experience difficult emo-tions such as anger or anxiety without being fused to them. Rather than *being* our irritability, we can *experience* being irritable.

We can also notice how thoughts color our experience of a situa-tion. We can watch distressing thoughts come and go, knowing they are creations of the mind. When we have a frightening thought about the future, we can stay grounded in the present and observe the pro-cess of thinking rather than getting lost in the content.

This facet of mindful awareness has tremendous implications for CBT. As we recenter in the present and observe our thoughts and feelings, we discover that these transient states are noisy. They often generate more heat than light, capturing our attention without tell-ing us anything meaningful. As such, we can treat them more lightly.

This is not to say that our thoughts and emotions don't matter or that we should always disregard them. They obviously can be valuable sources of information. But we'll be in a better position to evaluate them objectively when we practice decentering. As we become better acquainted with ourselves and how our minds work, we'll recognize when internal experiences are coming from an assured place of sensing and knowing and when they bear the signature of unhelpful fear or craving.

Look for an opportunity today to practice a different way of relating to a difficult emotion. If you're feeling angry, for example, become curious about the experience. Sense anger as a pattern of energy in the body. Study what's happening in the mind as if you're an "emotionologist." Open to whatever you find, and notice whether this approach changes your emotional experience.

Decentering allows us to see through countless unhelpful assumptions, which in turn guides our actions. Being is the ground for thinking and acting.

Think Mindfully

Sometimes we're unhappy not because things are going wrong but because we think they will. During the worst of my illness and depression, I often woke up with a feeling of dread in the pit of my stomach and anxious thoughts about the day ahead. "Today's not going to go well," I would tell myself. I didn't experience these thoughts as predictions my mind was making; I believed I was just acknowl-edging a fact about the day ahead. As a result, I expected the day to be marred by failure and disappointment.

When I began practicing mindful awareness of my thoughts, I started to recognize when my mind was telling scary stories. "Wait a second," I would think, "that's just a thought." In that moment, the grip of these negative thoughts would loosen, simply by my see-ing them for what they were. I could then consider alternatives, like "Maybe things will be okay today, like they are most days, even if there are difficulties." New thinking habits led to less anxiety and stress.

This way of applying the *think* approach was a powerful tool in my recovery; it was like trading a shovel for a snowblower to move mountains of heavy snow. It provided the leverage I needed to get out from under the weight of destructive thoughts. This technique has provided relief for countless people with conditions such as anxiety and depression. However, like any tool, it has limitations.

First, we might not believe the alternative. Maybe we recognize our thought as a thought but can't talk ourselves out of it. On a gut level, it just seems true. And second, even if we can convince ourselves there's a great day ahead, it could turn out to be filled with difficulty. In either case, our well-being is predicated on things working out the way we want, and we're at the mercy of factors we can't control.

With mindfulness as the center of CBT, more profound possi-bilities emerge—including a joy that doesn't depend on our cir-cumstances. If traditional cognitive therapy is a snowblower, the addition of mindfulness is like the warm spring sun that melts away the snow. In Think Act Be we can probe beliefs that go to the heart of our being.

GO DEEPER

Underneath the thoughts about my day ahead was a more fundamental belief: *my happiness depends on how things go today*. If things turn out well, I'll have a good day. If things go badly, I'll have a bad day. But another possibility emerged as I practiced mindful presence: maybe my well-being doesn't depend on outside circumstances.

Most of us find that this is a foreign way of thinking, and we might assume it's overly idealistic. "Of course my happiness depends on how things go," we tell ourselves. "Why would I be happy if bad things happen?"

Assuming that something outside ourselves is responsible for our ultimate well-being is a thinking error that I call "outsourcing happiness." This way of thinking is a nearly universal core belief that reaches into every area and moment of our lives, compelling us to see everything in terms of "for me or against me."

If we live by this core belief, we set a low ceiling on our well-being. At best, we'll experience isolated moments of peace, if we can even call it that—a fragile peace that could crumble at any moment. We'll also suffer in anticipation of things not working out in our favor as we look to the future. Few beliefs have a more powerful and pervasive effect on our happiness.

When I was falling deep into credit card debt because of my sickness and reduced work hours, I was afraid we were going to lose our house. I reassured myself that it was unlikely—we could always drain our retirement savings, or maybe family could help us out. Those assurances helped at times but not for very long. I knew there were no guarantees, and a few bad breaks could spell bankruptcy and foreclosure. Real peace was elusive.

Greater peace came from shifting the focus to my underlying belief. I began to tell myself, "Maybe maintaining life as I know it is not the basis for my peace and security." I realized that people with much less than I had managed to find joy. Although financial ruin would be a dramatic change, by no means would my life be over. Many people have lost their houses—including my own parents long after I left home—but their lives continued. It was a sad and painful process, but there was always the next thing to do.

With mindful awareness, we can see through the core beliefs on which we base our happiness. We find that joy is an option, even when life is difficult, as we connect with a part of ourselves that is untouched by all our measures of loss and gain. Our spirit is not defined by what's happening with our body or by our shifting roles; it is not diminished by a setback at work or dependent on the praise of others. We can let go of all the conditions we place on our contentment.

We don't have to convince ourselves that we should be happy to experience this kind of equanimity. It emerges when we simply inhabit the present moment.

Act Mindfully

A common reaction to the idea of mindful acceptance is that it implies passivity, complacency, or resignation: Accepting that I'm sick suggests I've given up hope of recovery. Accepting that my boss is difficult suggests I'll stay in an unpleasant job. Accepting that humans are destroying the planet implies that I'll stand by and watch it happen. In a similar way, cognitive techniques could be seen as a relatively passive way of making peace without changing our situation.

However, acceptance and clear thinking are the basis for effective action. When life isn't going the way we want it to, a third option is to take action and change the situation. When we receive reality as it is, we can then ask, "What is the appropriate response?" Accepting that we're sick can lead to seeking the right treatment. Accepting that our boss is difficult could lead to finding work elsewhere. Accepting climate change could lead to taking political action. Any constructive change we might make begins with accepting our situation for what it is.

An essential part of finding unconditional peace is focusing our attention and energy on what we're doing right now. Our thoughts tend to run ahead of us to the next thing, which projects our attention into the future and disconnects our minds from our bodies. Not surprisingly, we feel scattered and frantic, since we've fragmented ourselves across time and space. Disconnection is not a state of ease.

It's especially easy to feel frantic when we're constantly rushing from activity to activity. Constant *doing* tends to be self-perpetuating because it creates a vortex of stressful urgency that pulls us from one thing to the next. We often don't realize the stress we are taking on in our bodies, since our minds are somewhere else, focused on what's to come. Sometimes we need to slow down in order to catch up with ourselves—to reunite our minds and bodies in the present. As we come back to ourselves, we'll find we can act more purposefully, taking care of our tasks as we maintain connection with our center. The simple act of coming back to ourselves is inherently stress-reducing.

Notice today when you're rushing and feeling pressed for time. How much of your attention is in the moment, and how much is focused on what's ahead? How connected to your body do you

feel? **Realizing that our minds are rushing ahead of our bodies is the first step toward coming back to the present. In later chapters, we'll continue to work toward making friends with time.**

A crucial part of mindful action is giving up the illusion of control that was never really ours.

RELEASE CONTROL

In one of my first sessions with Jennifer, she told me, "I just want to be able to control everything . . . all the time." She knew this desire wasn't realistic, but still she clung to it. Jennifer desperately wanted to know she would be okay, and a loss of control implied a loss of security. Sadly, her fruitless efforts at ultimate control placed her in a battle against reality and were one of the main contributors to her constant anxiety.

I was furious that night in Ada's room that I couldn't make her go to sleep. But mindful acceptance helped me to acknowledge the limits of my control. When I received what life was offering in that moment, I surrendered my frustration and futile efforts. Releasing a false sense of control helped me reclaim the only true control I had—control of where I placed my attention.

The truth is, you're probably better off not being fully in control of your life. Think of the unfortunate outcomes that might have resulted if you'd been pulling all the levers. You never know if getting what you want may not turn out so well—like when winning the lottery ruins people's lives—or if "bad news" may end up working in your favor.

I spent countless nights and weekends laboring over grant applications when I was a junior faculty member. None of my research proposals were ever funded. I was disappointed with every rejection letter, but to be honest, I also breathed a sigh of relief. Getting a grant would have meant an additional three-to-five-year commitment to a job I found uninspiring, in a work environment that was far from ideal. When things didn't go "my way" with the grants, it actually became easier for me to move on to work I found much more rewarding.

Jennifer found relief from her anxiety by being willing to let go of her constant need for control. Again and again, she practiced the essence of the Stoic philosopher Epictetus's instruction: "Ask, 'Is this something that is, or is not, in my control?'" Most of the time for Jennifer, the answer was no. "And if it's not one of the things that you control," continued Epictetus, "be ready with the reaction, 'Then it's none of my concern.'"[2] As Jennifer released her stranglehold on outcomes she couldn't control, she began to open to life in all its uncertainty.

Embrace Uncertainty

When my wife was laboring in the hospital with our first child, I couldn't wait till Lucas was delivered safely into our arms. We had lost two pregnancies at the end of the first trimester, and I knew this final step that would bring our baby into the world was fraught with risks. "Finally I'll be able to relax, knowing he's safe," I thought. But then I realized there would always be the next unknown to worry about: sudden infant death syndrome, stairs, cars, driving. At no point would I be able to say with relief, "Whew, everything worked out."

One of the biggest blocks to our happiness is our relationship with uncertainty—and every outcome we care about is uncertain: our loved ones' health, our own health, the economy, how long we'll live. Even though we might accept this uncertainty on a rational level, a big part of us hates the feeling of life as a tightrope walk with the very real danger of falling. We want to know that our kids will be safe. We want to make sure we'll be able to pay the mortgage. We'd like some assurance that we'll live a good, long life and die a painless death.

More than anything, we want to know we're *going to be okay*. We can spend most of our lives chasing a mirage, believing we can reach some state at which life's inherent uncertainty disappears. But the more we chase this illusion, the more uneasy we feel.

Years ago, I worked with a successful corporate attorney named Bill. He was miserable in his work but felt that he had to keep pushing himself to safeguard a secure retirement. He'd planned to retire once he had saved $2 million, but then he worried that that amount might not last the rest of his life. So he doubled down. He was approaching $4 million in his retirement account when he came to see me.

At that point, Bill was desperate for some relief. He was constantly on edge, and the stress in his body and mind were palpable as we sat together. His health had suffered tremendously from the unrelenting stress, and his marriage would soon end in divorce. And yet he still didn't feel that he could step away from his work. Bill was nearly pleading with me in one of our sessions as he asked, "How much is enough, Seth? Four million? Five? Six?" He put his head in his hands. "Maybe I have enough now. I just don't know." It was a heartbreaking irony to witness—in trying to find peace of mind, he'd sacrificed exactly what he aimed to protect.

For Bill, mindfulness offered a way out. He began to extend acceptance not only to what was happening in the present but also to the unknown future. Maybe his savings would last until he died, or maybe the money would run out. He found that he could receive all of his experience, not knowing exactly what it would be. It was a huge exercise in trust, beyond believing that everything would work out the way he wanted it to; he'd lived long enough to know that wasn't the deal. He learned to trust that he would be okay even when everything was far from okay. He laughed as he described lying on a pier one moonless night, looking up at the countless stars and knowing he already had everything he needed. Bill found freedom through letting go.

Notice the next time you're plagued by "What if . . . ?" questions about an uncertain outcome, like "What if my sister is mad at me?" Trying to figure out the answer to these worries often leads to more worry and anxiety. Take a gentle breath in and out through your nose, and then experiment with a response like "She might be mad at me, and I'll need to deal with it if she is." Drop the struggle, and treat worried thoughts as an opportunity to face the unknown.

We can stop the hopeless pursuit of certainty and come to rest even with the knowledge that nothing we care about is guaranteed. We can open to the possibility that our balancing act will end in a glorious fall. We no longer have to wrangle with our worries. We don't have to take the bait when the mind asks, "What if something terrible happens?" The only reasonable response is, "It might. And if it does, I'll deal with it." By mindfully opening to that possibility, we let go of a pointless fight.

Know Yourself

One of the greatest benefits of settling into the present is discovering who you really are. Peace comes with this awareness, regardless of what comes our way, as we stand firm in our unshakable identity. When we connect with ourselves in the moment, we remember what has always been true about us beyond our superficial roles or traits. Life is much less terrifying when we know who we are.

You discover first and foremost where you exist—always in the present. That's where you can find yourself. Your body is here. Your mind is here. Your spirit is here. You can let go of fantasies of what may come because that imagined person in the future isn't you. You're the one doing the imagining. And the *you* worth taking care of is always right here.

Peace in the present also comes from knowing we're not individual islands but are intimately connected to all of creation. When we're fully in the moment with others, we're not just sharing space or exploring ideas together. A deeper part of ourselves is meeting the other. We sense in those moments that spiritual connection is part of our essence.

We also discover how strong we are when we abide in the present. Paradoxically, we may not realize the strength within us until we're feeling our weakest, as I discovered for myself. Your body may feel depleted. You might be mentally and emotionally drained. And yet there remains a deeper well of inexhaustible strength. Finding our center allows us to draw from that well and face whatever comes. We know we can tolerate pain. We know we can face our fears. We can feel the power in a single breath, connecting us to all of life in an unbroken chain of breath and being.

Through present-centered connection, we can realize the most fundamental truth about our nature: we are built of love. This isn't a hearts-and-roses type of love that's based in emotion and self-interest. It's a fierce and steady love that wants the best for all beings, ourselves included. This doesn't mean we always act or feel loving. There are occasions when we hate other people, ourselves, or life in general. But we can recognize those times as arising from disconnection to our true identity. Above all, you can know that you are loved, completely, exactly as you are. Yes, you. No matter the circumstances, lasting peace is found in that awareness.

Leaving the present costs us more than we know—our peace of mind, connection with ourselves, connection to others. With consistent practice, we can train our minds to come back to the present, where we'll discover how mindful awareness can shape our thoughts and actions. Thankfully, there are simple and well-tested techniques for being more fully in our lives, as we'll see in the next chapter.

PRACTICE MINDFUL AWARENESS

Mindfulness is an essential part of Think Act Be, as we saw in the previous chapter, offering a firm foundation for cognitive and behavioral practices. Sometimes mindful awareness catches us off guard, but we don't have to wait for it to show up uninvited. In this chapter, we'll explore simple and specific ways to cultivate greater mindfulness.

* * *

"I just don't think I'm good at mindfulness," Jon said during our first session together. He had started having unexplained bouts of near panic in his early forties, and a friend had suggested he try a popular meditation app, which Jon had downloaded and used off and on for a few weeks. "For some reason I was never very consistent with it," he told me. "The meditations helped me relax, and I wanted to practice every day, but more often than not I couldn't seem to find the time."

"Besides," he continued, "sometimes my anxiety spikes in meetings at work. It's not like I can pause and take ten minutes to focus on my breathing." For Jon, meditation seemed like a nice practice that didn't have much to do with his real life—which no doubt contributed to his reluctance to meditate regularly.

I found it easy to relate to Jon's story. I had known about mindfulness meditation for a long time and had even been applying it in my clinical work before I developed a consistent practice. I kept meaning to, and yet somehow I wouldn't get around to it. Finally, I started to meditate for ten or fifteen minutes first thing each morning after arriving at my office. I had some meaningful experiences during meditation, and I hoped that somehow those minutes of attending to my breath would carry through the rest of my day. But virtually none of my practice seemed to follow me when the ending bell chimed, and my mind would pick up right where it had left off. I felt as if I may as well have not meditated.

How can we practice mindfulness in a way that makes a difference in our daily lives? The first step is to recognize what gets in the way.

Recognize the Separate Self

Even when we like the idea of mindfulness and know of the benefits, most of us struggle to practice it regularly. We might plan to spend ten minutes each day in meditation but wind up meditating very rarely, if at all. Or we intend to find moments throughout the day to connect with our experience but then forget to ever be in the present. We can be genuinely perplexed by our behavior, especially if we find the practices helpful when we use them.

The key to understanding our behavior is to recognize the "separate self," often referred to as the "ego" in Eastern religions and philosophy (I'll use the terms interchangeably). Your separate self sees you as an individual entity, distinct from other people and from the world. This distinction feels good in a way as your ego creates a drama in which you're the leading character and carves out an existence you can call your own.

However, our separate self also senses that it's weak and vulnerable. No matter how much we have or what we accomplish, it knows that to be separate is to be small—like a drop of water apart from the ocean. The separate self thus fears annihilation and demands security; it's constantly on edge, looking ahead for danger and judging whether reality is "good" or "bad" to make sure it will be okay.

More than anything, the ego is concerned with self-preservation. That's why we so often struggle to open to the present. Mindful presence means letting go of the ego's insistence that the world obey its wishes. We say yes when the ego is saying no. We stay present when the ego wants to run ahead to the future. We release the false divisions the separate self creates between ourselves and others. We interrupt the ego's focus on "for me or against me." These moves cause us to be reborn into the present—while the separate self fears for its life.

As a result, it fights for our attention. One of its most powerful tactics is getting us to identify with it so we'll align with its goals. Most of the time, we identify with this version of ourselves, letting it shape our habitual actions and mindset. The ego tells you that "you" don't want to meditate or say yes to your circumstances. In reality, the resistance is not coming from *you* but from a part of you that often controls your mind. Knowing where the resistance is coming from makes it easier to let it go.

Nevertheless, most of us find that knowledge and desire alone are not enough to overcome the ego's tenacious resistance to mindful presence; we need effective techniques to develop a consistent practice. This need took on extra urgency for me as I began my own cognitive behavioral therapy (CBT) at the low point in my depression. I had felt the life-changing power of mindful connection to the deeper parts of my being when I came to the end of my separate self. I knew implicitly that fostering that deep connection would be central to my recovery.

Cognitive and behavioral practices offer the leverage we need to transform our desire for mindful awareness into meaningful change. Thinking and acting are mutually reinforcing, and together they support being. Being, in turn, guides our thoughts and actions in a self-sustaining cycle (see figure 3). Think Act Be offers a way to bring mindfulness practices into our everyday lives.

Figure 3

Let's take a look at how we can practice mindful presence on purpose. We'll begin with how to train our minds.

Think: Question Beliefs About Mindfulness Practice

Seated meditation provides a concentrated dose of intentional focus on something in the present—most often the breath.[1] You just notice the sensations of breathing, such as the gentle rise and fall of your abdomen, each time you breathe in and out. When you realize your attention has drifted and you've been lost in thought, you gently return your awareness to the breath.

I've moved away from formal meditation at various times in my mindfulness practice, with the intention to focus on mindfulness in my day-to-day activities. But, like most people I've worked with, I've found that there's nothing like meditation for stepping out of my habitual ways of thinking and doing and connecting with my present-moment experience. Trying to be present without practicing meditation is like trying to learn your part in the orchestra while the rest of the musicians are playing different pieces.

Meditation offers the space we need for seeing the workings of the ego and mind, with the only interruptions coming from our thoughts. Regular practice helps us to develop the habit of being aware of what the mind is up to—when it's in the present, when it's in the future or the past, and how open it is to what's happening right now. That awareness is the foundation for the Think Act Be approach, helping us to work with our thoughts more effectively and to notice the effects of our actions.

I invite you to try sitting in silent meditation, especially if you've done it before and didn't like it. You may find that a different way of approaching the practice makes it more enjoyable and more relevant to your life. As we'll see, many of the obstacles to practicing meditation come from unhelpful beliefs and expectations.

LET IT BE EASY

I found that one of the hardest things about mindfulness practice was to let go of effort. For years, I treated meditation as one more item on my daily to-do list—*mindfulness . . . check*. This task orientation turned meditation into a chore that I could do incorrectly. Without deliberate attention, we'll end up folding meditation and other mindfulness practices into our habitual ego-driven way of "getting things done." But mindful presence is about doing less, not more. It's the simplest thing you will ever experience.

When your mindfulness practice feels like a job or you notice thoughts like "I'm not doing this right," see whether it's possible to do less. Simply receive awareness, just as the ears receive sound without any work on your part. Rather than striving to be mindful, release into it.

FIND CONNECTION

I used to find that sitting meditation was a lonely experience, since it was only about *my* effort and *my* focus. I felt like a lighthouse in the night, coldly directing the spotlight of my attention on inert objects that gave no reply: "Now I'm noticing this. Now this." But even though we often meditate in solitude, it's least of all about our individual effort. Mindfulness is a relationship.

We find ease in our mindfulness practice not by aiming our attention but by relating to our experience. Everything within us and around us and everyone we meet is calling for connection. My experience on the couch was a connection with my spirit. My crib-side experience was a connection with my baby. Cycling up hills was a connection with my body and my bike. Even the seat you're sitting on and the book you're reading call for your awareness and connection—to feel them, to know them, to experience them.

We find as we practice mindful awareness that we are truly beings of intimate connection. We discover that connection not by adding more effort and activity to our lives but by letting go of the illusion of separateness that arises from the ego. When that veil slips away, relationship inevitably emerges. At its most elemental, mindfulness is a relationship with the unfiltered experience of being.

WATCH OUT FOR "SHOULD"

Many of us assume that practicing mindfulness means we have to meditate. But telling ourselves we "have to" or "should" be mindful is an unhelpful starting point for being present, since it creates a judgmental mindset.

As you explore your own mindfulness practice, watch out for thoughts like "I should meditate" that turn mindfulness practice into a guilt-filled obligation. Try out alternative ways of thinking, like "I'm learning to practice being present" or "I plan to meditate more often."

MAKE PEACE WITH HAVING THOUGHTS

One of the biggest stumbling blocks in meditation practice is the belief that we should be able to clear our minds. Most of the people I've treated who tried meditation bemoaned how bad they were at "making their thoughts stop." But that's what minds do—they generate a nonstop stream of thoughts, images, memories, and fears. If we try to stop the stream, our mind just gets louder and more insistent.

Getting rid of thoughts is not the goal; instead, just observe what the mind is doing. You'll still be aware of thoughts that come and go while you focus on your breathing, but they can exist in the periphery of your awareness. It's as if you're sitting in the sand watching the ocean waves roll in and out; you notice birds coming and going, but you can keep your focus on the waves. Sometimes you'll find that your attention has flown away with a flock of thoughts. When you realize your awareness has drifted, gently guide it back to the waves of breath. Keep this cycle in mind as you're meditating: you're aware of your breath, you lose that awareness, and then you find it again (see figure 4).

You might also notice a barrage of thoughts at the beginning of your meditation sessions. Suddenly you'll remember an email you need to write, and several tasks you *must not forget to do*, and other thoughts that clamor for your attention. Keep bringing your focus back to the waves as this cloud of birds passes through the scene in front of you.

LET GO OF SELF-CRITICISM

Ironically, we often make harsh judgments about how we're meditating. I once asked my patient Victoria how her meditation practice was going. "Ugh, I'm *terrible* at it!" she replied. Knowing the process of meditation—awareness, drift, return—helped her to

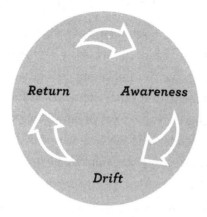

Figure 4

let go of the false belief that she was bad at meditating. Shifting away from self-criticism in her meditation practice helped her to become more compassionate toward herself in the rest of her life.

More than anything, remember that you're doing just fine, no matter what happens during meditation. You're not doing it wrong. You're not a bad meditator.

Take a mindful stance toward your mindfulness practice itself by accepting however your practice turns out. If meditation feels relaxing, enjoy it. If it's difficult and stressful, that's how it was this time. If you fall asleep, you obviously need the rest. If you spend a whole session mentally adding items to your to-do list, that happens. And if you judge yourself, that's okay, too! Seriously, you can't mess this up.

ALLOW FOR BOREDOM

The simplicity of meditation isn't always a welcome experience. Feeling bored is one of the most common reactions to meditation. Meditation *is* boring on one level, compared with the constant stream of stimuli that we're used to. But we don't have to let boredom stop us from meditating. We can notice our thoughts about how bored we are, just as we notice any other thoughts, without getting lost in them. Try exploring the experience of boredom—being in relationship with your boredom—and see what happens. *Being* bored is boring, but actually *experiencing* boredom can be quite interesting.

Experiment with boredom by sitting and doing nothing—no books, no screens, no work. Notice how your mind and body react. Does it feel like a welcome pause? What emotions arise? Is there a strong urge to do something productive? What enters your awareness when there's no activity to fill it?

VALUE BEING

Most of us have busy schedules and feel as if there aren't enough hours in the day. In this context, it's easy to believe that meditation is a waste of precious time. This judgment makes sense if we equate meditation with "doing nothing" and assume we'll have nothing to show for those minutes spent in meditation. But this belief betrays the ego's obsession with *doing* and *achieving* and its undervaluing of *being*.

The most effective antidote to this belief is to do a few sessions of meditation. Taste and see. There's no substitute for firsthand experience of the practice, in which we discover that connection in the present can be inherently enjoyable.

Open yourself to the possibility that your experience matters, quite apart from what you have to show for it. Value *being* for its

own sake. **Approach meditation as an opportunity to observe ev-erything that happens rather than as a task with a goal in mind.**

RELEASE THE OUTCOME

My patient Jon had stopped meditating when he found that it didn't get rid of his anxiety. Many of us expect that meditation practice will get rid of the less desirable parts of our lives, such as stress, anxiety, and other difficult emotions. Sometimes that happens, given the calming nature of mindfulness practice. But feeling more comfortable is not the purpose of openhearted presence—and it is less likely to happen if we make it the goal.

Expecting certain outcomes from our mindfulness practice will get in the way of our experience. Expectations lead to evaluation—Am I relaxing? Is my mind quieting? Am I having a mystical experience?—which pulls us out of the present. Instead, we can approach each moment of meditation as if it's the first time we've experienced it, without preconceptions or goals.

Mindful awareness is about relationship. In my work with Jon, he practiced a new way of relating to his anxiety. When the anxious spells came, he opened to the experience with curiosity rather than resistance. Instead of telling himself, "I can't stand this!" and "I have to make it stop," he said, "Let me see what this is like. What's happening in my body? How does the anxiety shift over time?" Jon discovered not only that he suffered much less but also that the attacks came less often.

Mindfulness changes our relationship with our experience, not necessarily the experience itself. When I opened to my experience in the nursery that night, mindful awareness didn't make Ada fall asleep.

When I was cycling, it didn't flatten the hills. During my illness, it didn't cure my chronic insomnia. But I found peace nonetheless—the peace that comes when we open to our difficult experiences.

Clear thinking about how to practice mindfulness prepares us for action.

Act: Practice Mindfulness

When you're starting a meditation practice, these behavioral principles provide leverage.

START SMALL

Apply the *act* principle of making it easy to do what you want to do. Start out gradually if you're new to mindfulness practice. Five minutes of meditation can be a great beginning—long enough to get a feel for what it's like and to tap into the benefits but not so long that it feels like a prison sentence.[2] If you love it, you can always do more.

Spend a few minutes in seated meditation. Set a gentle alarm to let you know when time is up, or follow along with a brief guided meditation. You can even do it now if you like, before you read the rest of this chapter.

MAKE IT ENJOYABLE

Choose meditation practices that you like to do. As with physical exercise, the best practice is the one we'll use consistently, and that means the one we find rewarding. My baseline practice is to sit in silence on a yoga block and follow my breath. When I was physically exhausted, I followed along with a guided meditation while I lay on my back. When I wanted to develop greater awareness of my body, I did guided body scans.

As you explore the mindfulness exercises throughout this book, move toward the ones that resonate with you, and make adjustments to suit your changing needs. There's no wrong way or better way to come into contact with ourselves and the world.

PUT IT IN YOUR CALENDAR

Other activities have always crowded out my meditation practice when I've let them. As with anything I want to commit to, I now dedicate a specific time for it (first thing in the morning). If you need a reminder, just set an alarm for yourself.

PRACTICE NOW

I went through so many days with the full intention to practice meditation, which somehow never materialized. I fully planned to meditate, but *now* just never seemed like quite the right time. When there's always one more thing leading you to put off mindful presence, suspect the subtle workings of the ego, telling you "Sure—just not right now." When we realize that "later" often turns into "never," we can see through this false belief and choose to practice now.

MEDIDATE

Accountability can be a helpful tool in making meditation a regular practice. A meditation date, or "medidate," is a simple technique that can help with consistency. Partner with a loved one and plan to meditate together. You might find that it's a surprisingly powerful way to strengthen your bond.

EXTEND MINDFUL AWARENESS

In principle, meditation prepares us for a fuller experience of our daily lives—the greatest promise of practicing mindfulness. And yet like many others, I found it hard to connect my morning meditation sessions with the rest of my day. It's easy to understand why many of us give up on meditation if we find no obvious benefit in our daily lives.

Meditation isn't the end point of mindful awareness. Mindfulness is about *all the time*, not just the few minutes we spend silently connecting with our body and breath. Even a full hour of meditation each day would leave fifteen or sixteen hours in our nonmindful default mode. We need a way to translate our focused practice into skills we can apply to the rest of our day.

One of the most helpful approaches is to deliberately extend our meditation into the moments immediately after. Following a breath-focused meditation, we set an intention to maintain mindful awareness as we open our eyes, stand, and move toward our next activity.

MEDITATE IN MOTION

We can also use moving meditations, such as yoga and tai chi, as an intermediate step between sitting in stillness and staying present in our daily activities. The repetition of the movements in these disciplines offers a familiarity that parallels the simplicity of being with the breath. These more active forms of structured mindfulness practice in turn can be an effective transition from formal practice to the rest of our lives.

Here, I'll focus on yoga because it's the moving meditation I'm most familiar with. To be honest, I'd been practicing yoga for years without understanding why it was considered a mindfulness practice or how it could influence other parts of my day. The few minutes of meditation that often opened a yoga class were obviously connected to mindfulness, as was lying in stillness at the end of class. But all that movement in between just seemed like exercise, with no relevance to my mind or emotions off the mat.

Several years ago, it finally clicked for me as I stood with the rest of the class in a pose called Warrior II: left leg bent in front of us, right leg extended behind us, arms reaching forward and backward at shoulder level. Normally I was focused on the physical challenge of trying to do the poses. But this time, our instructor, Mark, invited us to feel in our hands the *intention* to move before we transitioned to the next pose.

That simple suggestion suddenly made me aware of the mindful part of yoga that I'd been missing, which had to do with how I was relating to my body. I had always felt I was *deploying* my movements to accomplish Warrior II and all the other postures. I commanded; my body obeyed. But underlying the motions was an awareness of a *relationship*—a joining of body and mind. I'd been focused on the messages I was sending out, without receiving the communication that was coming back.

Yoga is a mindful practice when it yokes mind and body in an intimate exchange. When we arrive on our yoga mat, we can begin by setting an intention to be present for the practice. Starting with a couple minutes of seated meditation can help us settle into mindful awareness as we establish a connection with our body and our breath.

We can maintain the relationship between mind and body as we start the moving part of the class. The movements from one pose to the next are paired with the breath, which helps keep our awareness grounded in the moment and in the body. When our attention wanders (as it will) to other things, we guide it back to what's happening within the four corners of our yoga mat: physical sensations, the air on our skin, our breath.

There are also countless opportunities in yoga to practice letting go of resistance and opening to our experience, which we can carry with us into the rest of our day. We practice acceptance when we stay through the manageable discomfort of a challenging pose, which can help us to grow both physically and mentally. We accept that our practice probably doesn't look like the instructor's as we embrace our physical limitations.

As with meditation, notice the mind's stories that could interfere with your practice. Common thoughts include versions of "this is too hard," which can lead us to step out of a pose. There are times when ending or modifying a pose is the most compassionate thing we can do for ourselves. But our core belief that discomfort is intolerable often underlies these self-limiting thoughts. As we push the boundaries of what we're able to do, we build not just our physical strength but also our capacity to tolerate discomfort in the service of something we care about.

If you practice yoga, question the assumption that you "can't stand" a challenging sensation in a pose. Acknowledge the discomfort without letting it drive your actions. Exercise your cu-

riosity, exploring the uncomfortable sensations you feel instead of trying to ignore them or push them away. Like boredom, physical discomfort is less unpleasant when you study it more closely.

Personally, I find that yoga provides the easiest transition into unstructured mindfulness. When our class (or online yoga video) ends, we can just keep connecting to our bodies in the present—noticing what it's like to move, feeling the breath in the body as we roll up our yoga mat, drive home, or make breakfast. It doesn't matter how long we maintain this connection, whether for a single breath or for the entire day. Any extension of mindful awareness into our walking-talking-moving-doing lives will be beneficial and will set us up for more helpful thoughts and actions.

BRING MINDFULNESS WITH YOU

Most of our opportunities to be in the moment are in our day-to-day activities, from the moment we open our eyes in the morning until we drift off to sleep at night. The invitation to be fully in our lives is always extended. But we can easily go days, or years, scarcely remembering that mindful presence is available.

Mindful awareness is like a silently flowing river that's always present but often forgotten. Sometimes we'll happen upon it, as I did one evening in my kitchen when I heard my kids playing in the basement. I was suddenly overwhelmed by the awareness of how much love and beauty filled my life, which moments before had been obscured by other thoughts and concerns. "How was this here the whole time?" I wondered. "How did I miss it?" It was as if I'd spent endless hours looking for clear, cold water to quench the thirst of my vague discontent when the river I was searching for was right here all along.

That's the beauty of mindfulness—it's always right here, as close as anything, and all we have to do is step in. No matter how long we've been away from it, we can always return in an instant. The challenge is remembering to show up—which is where the behavioral component of CBT can be so helpful. We can set reminders to return to the present, or we can leave ourselves notes: *Pay attention to brushing your teeth. Three mindful breaths. Open to what's happening.*

> **Look for specific activities during which you can practice mindful awareness—for example, chopping vegetables or taking a shower. Committing to circumscribed practice is easier to remember than a generic plan to be mindful "all the time."**

It's also helpful to plan moments of deliberate reconnection throughout the day to bring you back to your experience. They might include three minutes of quiet self-reflection upon awakening, short pauses from time to time to focus on the breath, and a brief meditation before bedtime.

> **One of my favorite exercises is a Breath Minute: Time how many slow, conscious breaths you take in sixty seconds. Then, as often as you need to throughout the day, pause and take that number of breaths. You'll know it takes only a minute and you won't have to use a timer.**

Mealtime is another good opportunity to come back to mindful awareness. A single breath with our full attention can act as a reset button and prepare us to approach the meal from a calmer and more centered place. I often take three breaths before each meal: one as I check in with what I'm feeling, one as I take in my surroundings and those I'm eating with, and one as I behold the plate of food in front of me.

As you invite mindful presence, you don't have to force anything or try to make it feel "spiritual." Keep it very ordinary and uncomplicated, and just notice what's happening: See what you're seeing. Hear what you're hearing. Take in colors and textures around you. You can tune inward, too, seeing what emotions are present and watching what your mind is up to. This can all happen in real time as you go about your activities. You can try it with cooking, cleaning, walking, bathing—anything at all, including reading this book.

As you pay attention, open to anything that comes your way. Receive it. Proactively say yes to it all. Release the constant drive to improve your situation. Settle into it instead: "This is what's happening. This is my reality." That doesn't mean you don't fix something that's wrong or tell someone no. Just stay open to all of the experience as it's happening, even the uncomfortable parts.

If you're having a hard time following through on your efforts to practice mindfulness, see what it's like to integrate the three parts of Think Act Be in order to move toward your intention.

First, *be*: Close your eyes and take a slow breath in and out. Feel any sensations in your body, such as your feet on the floor or a tingling in your hands. Become aware of any emotions you're feeling.

Second, *think*: Notice whether any thoughts are getting in

the way, such as "Meditation is a waste of time" or "It probably won't help anyway." Perhaps the mind's story is accurate, but see if it might be a bit distorted. Is there an alternative way of thinking that could be more valid? For example, "Each time I meditate, I'm giving my mind a chance to rest in the present."

Finally, *act*: Is there a small step you could take toward your goal? Maybe you could open your meditation app or roll out your yoga mat. Even if you go no further, this exercise in itself is a practice in mindful awareness.

Remember the Rewards

One of the most effective ways to support consistent mindfulness practice is to remember the many benefits we can look forward to. What follows are descriptions of what I and others have found when practicing mindful presence. Although expectations for a single practice session may be unhelpful, it does help to remember that there's a purpose to our practice. As always, beware of turning these descriptions into goals, since pursuing any state of mind takes us out of our present-moment experience. More than anything, rely on your subjective experience to guide your understanding.

RELAX

After I guide a patient through meditation for the first time, I always ask what it was like for them. Like most of my patients, Victoria said it was "relaxing," which is part of why meditation is often a core part of stress-reduction programs. There were multiple reasons why Victoria found that sitting quietly and focusing on the present was so calming.

First, meditation led her to breathe more slowly, and the pace of the breath is closely connected to the autonomic nervous system. Slower breathing turned on her parasympathetic system—the counterpart to the sympathetic system (fight/flight/freeze)—which calmed her mind and body.

Those few minutes of meditation also helped her to become aware of unnecessary tension she'd been holding, such as raising her shoulders in a half shrug and tightening her belly. Just noticing the tension allowed her to release it.

Beyond these direct physiological effects, meditation also soothed Victoria's mind. When her mind left the present, there was no limit to the problems it could imagine. As it looked ahead to the future, it saw issues with her health, her work, her family, her finances. Victoria found it was extremely stressful to try to work out all those problems in advance, like trying to keep in mind every intersection and turn she would have to negotiate on a long road trip. Her mind was also prone to worry about issues in the miles behind her when she dwelled on things in the past.

When Victoria gathered up her attention from the past and future, she let go of endless opportunities for worry and regret. She discovered that most of the time in this very moment, things were pretty much okay. Even if she was dealing with a problem, like an argument at work or bad news from the doctor, she learned that she was exquisitely well adapted to handle one thing at a time—like negotiating the stretch of highway right in front of her—which was all she ever had to do. As she abided in the present, Victoria discovered to her great relief that it was always manageable and that the seemingly unmanageable future became individual moments that she could handle.

AWAKEN

We spend most of our time lost in a dense web of dreamlike thoughts—memories, regrets, plans, fears, imaginary arguments and devastating comebacks. The thoughts feel so real that we mistake them for reality. But the content of those thoughts is a fantasy that's playing out only in our minds. Meanwhile, the bed of our existence is forgotten.

When we start paying more attention to the present, there's often a feeling of waking up as we come to our senses. We might notice there's a sky up there, and right now it's a particular shade of blue. We knew it was there, but it was so far from our conscious awareness that it may as well not have existed. We might feel how solid the ground is; it was always beneath our feet, but we rarely realized it.

We might wake up to sounds, too. As I write this, I become aware of all the sounds around me that had faded into a background hum: kids playing next door, cicadas chirping, a bird's call, an edge trimmer in the distance. We can also notice things about ourselves that we had missed, like that we're hungry or our toe hurts. We can see things we had overlooked altogether, as when we notice for the first time a house we must have driven past a thousand times.

None of this means that we *should* be continuously aware of everything that's happening, or that we're doing something wrong if we're not noticing every stimulus. It's good to be able to filter out noise when we're working. We need a kind of tunnel vision when we're puzzling through a difficult problem. There are obvious advantages to noticing only what's relevant when we're driving.

Nevertheless, it's easy to regret how much we've been missing. My introduction to mindful presence came largely through a little book called *Shambhala* that opened my eyes to much of my life that I'd been ignoring.[3] I was so focused on work and whatever passing thoughts were grabbing my attention that I'd been overlooking most of my life—including my three kids, whom I love dearly but had taken for granted in my semicomatose state. Waking up was amazing; I felt like Ebenezer Scrooge on Christmas morning with a second shot at life. It was terrifying to realize that I could have lived out my years without ever really waking up.

No matter what we start to notice that we've been missing, we can always look deeper and see more. As we continue to practice mindfulness, we find doors opening to new levels of experience, even ones we hadn't known existed.

DISCOVER

Eating a raisin can be a surprisingly emotional experience, as I learned in my therapy office. Valerie was the first of many people I led through the Raisin Exercise, a common practice in mindfulness-based CBT.[4] She had come to see me for help in dealing with the incredible stress of practicing law full-time while raising a three-year-old and a one-year-old.

As I placed a raisin on a napkin and handed it to Valerie, I asked her to imagine that she'd never encountered one of these little objects before. I then led her through a series of invitations to pay attention to the raisin more deeply than usual, such as seeing the variations in its color and being aware of her hand as it placed the raisin in her mouth.

To be honest, it seemed a bit silly to be paying special attention to this humble object. When we reached the end of the exercise, I paused for a moment and then asked Valerie what it had been like for her. As she began to respond, she wiped tears from her eyes. "I'm kind of embarrassed to say this," she began, "but I felt like I was going to cry." I had found the exercise strangely moving, too, so I wasn't surprised.

There's something poignant about noticing what we've always overlooked when we pay attention so deliberately. When we're fully awake to any of our experiences, we can see that whatever is happening right in this moment is incredibly interesting. Even the line where a wall meets the ceiling can be mesmerizing.

Practice perceiving parts of your everyday experiences that you usually miss: the coolness of water on your hands; the grain of a wooden table under your palms; the sensation of stretching to reach an upper shelf; the grooves on a raisin. Does your relationship to your ordinary life feel any different as you invite greater awareness?

When we start paying attention on purpose, the senses can take on a very sharp quality. We might see individual hairs on the back of our hand, feel our clothing on our skin, notice the variations of color in a person's eyes. And if you go a bit deeper, you notice that awareness is a continuous stream of noticing: Now this sensation. Now this emotion. Now this thought. Now this sound.

There's something that seems true about noticing at this level. It can start to feel miraculous to have any experience at all. None of what we take for granted was ever guaranteed, even the fact of existence itself, much less our own improbable appearance in it. We can start to feel grateful for everything, even the less desirable parts of our experience.

EXPAND

The resolute atheist Sam Harris is unabashed when he uses the word "spiritual" to describe "the efforts people make, through meditation, psychedelics, or other means, to fully bring their minds into the present or to induce nonordinary states of consciousness."[5] Even if we don't approach meditation as a spiritual exercise, certain forms of consciousness lead to mystical experiences.

These experiences often emerge when we settle into a single-pointed awareness. Meditation fosters the decentering we explored in chapter 4 as we witness all that we're aware of without being caught up in any of it. We simply note everything that shows up in our consciousness—thoughts, sounds, physical sensations—while our contact with what we're aware of stays light. We're finding connection without attachment, as if we were letting birds briefly alight on our open palms without grabbing hold of them.

At some point, awareness might begin to look back at itself, and you'll start to notice the very act of noticing. This is where things can get a little trippy; I find that words are inadequate to capture even the fleeting experiences I've had of this mode of consciousness. We approach a singularity of awareness, as if moving through a cave that grows narrower and narrower. Just when we think the passage will collapse into nothingness, we emerge into an enormous diamond-filled chamber, in which we are one of the diamonds. For just a moment any sense of "I" disappears, and we are just as much what we are observing as we are the observer—as if we are there and here, and nowhere, at the same time. The self seems to disappear or to be subsumed into a wider being. (This was my experience in my dream of death: I expected my consciousness to be obliterated but instead found eternal communion among the stars.)

William James noted that these "mystical moments" of "cosmic consciousness" often make a "deep impression . . . on those who experience them."[6] We feel that contact with something greater than ourselves—something deeply true—has changed us. Even though we're no longer in the same frame of mind and heart as during meditation, we've brought something back from our psychic spelunking.

Profound experiences of *being* change how we think and how we act. We find it's easier to see through the mind's illusions that we mistook for reality, like the assumption that we're deficient or the drive to acquire more. We want to align ourselves with that deeper reality and focus on love and connection. "*This* is what matters," we realize, "and *this* is how I want to live." Our head and hands support the work of the heart, and the Think Act Be cycle continues.

Every part of our lives is calling us to greater connection. This quality of connection feels right, like coming home. We find connection through the simplicity of being, which guides how we think and act. Our thoughts and actions in turn sustain connection. You can start to experience more of this connection wherever you are, by inhabiting the three-dimensional world.

6

CONNECT WITH
YOUR WORLD

We know from ancient Buddhist, Hindu, and Stoic texts that being caught up in thoughts and making evaluations is nothing new. But digital devices present a special challenge to connection with our inner voice, our sensory experience, and the people in our lives. Our screens continually grab our attention and offer immediate escape from the possibility of boredom or discomfort. Many of the most successful apps are designed to amplify the workings of the ego, reinforcing our sense of separateness and our judgment of everything as "for us" or "against us" (such as the number of people who "like" our social media posts).

This chapter offers ways to experience greater connection in a digitally saturated world, using the Think Act Be techniques we've explored. By limiting the sprawl of technology in our lives, we create more opportunities to be truly present for ourselves and others.

* * *

The emergency room doctor stuck her head into our curtained-off examining room. "Your daughter's CT scan looked fine. We're just waiting for the blood work to come back before we discharge her." I breathed a sigh of relief. Earlier that night, Ada had been having intense stomach pains, and our pediatrician had recommended a trip to the ER.

By now, it was eleven o'clock—well past both of our bedtimes—and the pain had subsided enough that Ada was able to sleep. When the doctor left, I instinctively reached for my phone to fill the time and prevent boredom while nothing was happening. But that night, for some reason, I said yes to a quiet voice inside that said, ". . . Or you could just be right here."

So I put my phone away and sat and looked at my seven-year-old, still pale but sleeping peacefully. Suddenly I felt incredibly thankful for her and her health and for having doctors who are available to us twenty-four hours per day, every day of the year. I didn't care that this would cost me twelve hundred dollars out of pocket. I didn't mind being awake so late.

My daughter was safely asleep in the hospital bed, and every indication was that she was okay. I felt overwhelmed with gratitude for her, for this hospital just five minutes from our house, for the skilled nurses and doctors who provided loving care. I was thankful for my own life, too, given my health struggles of the past couple of years. The "nothing" that was happening here was actually everything.

See what it's like to resist the urge to immediately take out your phone the next time you're waiting for something—for example, when you are standing in line at the grocery store. Open

your awareness to what's happening. Who or what do you see? What do you feel inside? Let the experience unfold on its own rather than trying to force any particular outcome.

None of that experience was waiting in my phone. If I'd taken it out, it wouldn't have been the end of the world by any means, but my world would have been diminished in ways I never would have known.

Make Peace

Like most of us, I have a complex relationship with my smartphone. It's hard to navigate our high-tech world without one of these mini-computers and impossible to deny the incredible usefulness of many recent digital innovations. At the same time, there are costs to spending too much time with our screens, which became clearer to me as I sat in my living room one night and planned my self-guided cognitive behavioral therapy (CBT).

I knew I needed to start with behavioral activation, which had lifted many of my patients out of deep depression. This simple yet powerful treatment involves doing more things that are rewarding either because they are enjoyable or because they feel important to us (or both). I was working on step one: taking an inventory of my current daily activities and rating each one for enjoyment and importance.

While I was working, my eyes fell on my iPhone, lying on the coffee table in front of me. As I picked it up out of habit to check my email, it struck me how much time I was spending on it: social media, email, news, sports, podcasts, YouTube, messaging. In some ways it was great, offering many conveniences and endless sources of entertainment.

But my activity assessment forced me to ask myself how rewarding the time I spent on my phone was. I had to be honest—*not very*. Most of what I used it for didn't feel important and wasn't even enjoyable. Yet I was drawn to it countless times throughout the day. A verse from the book of Isaiah came to mind: "Why spend money on what is not bread, and your labor on what does not satisfy? Listen, listen to me, and eat what is good, and you will delight in the richest of fare."[1]

My screen time wasn't exactly the richest of fare. I was more often checking email multiple times per hour for new messages or reflexively opening social media when I felt bored. The downsides were obvious: I wanted to stay present in my life, but I was constantly drawn into the alternate reality of my phone and was less engaged with my family than I wanted to be. More than anything, it felt as if my mind were not my own. I wasn't thrilled about changing my relationship with technology, but I knew deep down that it would be a necessary part of building the life that I wanted.

Start to notice how often you're looking at your phone or other screens, especially when you don't have to be. Release any self-judgment and just observe your patterns with curiosity. What kinds of emotions go along with your screen time? Do certain situations or feelings tend to push you toward picking up your phone?

Given the deliberately addictive nature of our devices, many of us struggle to use them without letting them dominate our lives. If we're addicted to substances like alcohol or cocaine, completely avoiding them makes sense because our life is only improved by swearing them off. But for most of us, abstinence isn't a practical solution to the problem of screen addiction. As with problematic eating, our challenge is to find moderation.

Simply believing we "should" spend less time on our phones tends to yield more guilt than meaningful change. Commanding ourselves to spend less time on our phones relies on our willpower, which we know isn't a reliable strategy—especially when the payoff structure works against us. If I'm thinking of spending more time on Facebook, I might hesitate when I remember that too much social media consumption can lower life satisfaction.[2] But long-term costs are much less compelling than short-term gains. The possibility of lower life satisfaction in the future feels as remote as the distant specter of lung cancer to a person who smokes; it's no match for the immediate reward that social media provides, like the instant pleasure from lighting up.

My past efforts to reduce my screen time hadn't been very successful. I might have curtailed my use for a while, but gradually my resolve would wear away and I would be back where I'd started. I would need a different approach this time—one that went beyond the simple tricks and willpower I'd tried in the past. I would need the full suite of Think Act Be: head, hands, and heart.

Be: Fall in Love

Emotions such as guilt and fear aren't sufficient to help us resist the pull of our screens. Love, on the other hand, is a robust and enduring source of motivation. We find love not in severing ties with technology but in connecting with the world around us. That's the message I've heard again and again from my patients—reducing screen time is not about giving something up but about finding something better. The benefits of mindful presence are typically subtler than the dose of dopamine we get from our screens, but they are powerful in their own way.

Beneath the demanding voices of email and social media, there's a softer call that beckons us to stay close to our experience. It's not a shaming or scolding voice. It doesn't threaten us with bad things that will happen if we ignore it: "If you pick up that phone, you're gonna get depressed. I can't believe you're checking email again." Instead, it opens its arms and says, "Listen to me, and eat what is good . . ."

I've ignored that voice countless times. "Sure, another time," I reply. "Right now, I just want to watch one more comedy clip. This one looks really funny." But when we follow that gentle voice, reality can be so much more than we're expecting, as I discovered that night in the emergency room.

Any direct contact with the world, from unloading the dishwasher to taking out the trash, can draw us into an expansive experience. As we've seen in previous chapters, we need only bring more of our attention to exactly what we're doing. However, we can find a unique connection when we spend time outdoors.

STEP OUTSIDE

Weekend afternoons unfold somewhat predictably in our home. After lunch, our kids play together, and at first all is well. Then, gradually, the harmony turns into conflict, with name-calling and yelling and hurt feelings. At some point, my wife and I announce that we're going on a family hike. "Noooooo!" the kids reply. At least they agree on something.

But we rally the troops and head out, and the moment we set foot on the trail, everyone's spirits lift. Our kids run ahead and play together, their earlier squabbles a distant memory. My wife and I feel calmer and more grounded. It's as if our family has taken a collective deep breath together just by stepping into the woods.

Research findings from a growing number of studies confirm what we know from direct experience—being in natural settings is inherently uplifting. Greater access to green spaces is associated with less anxiety and depression.[3] These benefits are even more pronounced when we have a lot of stress in our lives[4] because spending time in green space has measurable effects on physiological variables, including lower levels of stress hormones, lower heart rate, and lower blood pressure.[5]

I instinctively sought out time in nature when I was struggling through my illness. For years I had walked to and from work along the suburban streets that led to my office, but I changed my commute so that the nature trail of a local college bookended my workdays. I felt I was gathering strength for the day ahead as I walked the path in the morning, the towering poplars like old friends offering stability through a chaotic time. No matter what challenges I dealt with, my walk home through the woods at the end of the day was soothing.

Consider spending a little more time outside today than you normally would. Find any excuse to be outdoors even briefly—a short walk, dining al fresco, opening the mail on the porch.

Gently direct your awareness to your surroundings, taking in the sky, the light, and the plants. Notice how it feels to connect with the natural world.[6]

My patients have often told me about their own experiences of the outdoors, which they tend to find hard to put into words, probably because the connection we find there transcends language. They often describe a sense of healing when they're outside—something inherently life-giving. The language of poetry seems best suited to capture our relationship with the natural world; the poet Mary Oliver described an intimate connection between ourselves and even the air we breathe:

> *And you will hear the air itself, like a beloved, whisper:*
> *oh, let me, for a while longer, enter the two*
> *beautiful bodies of your lungs.*[7]

The plant ecologist Robin Wall Kimmerer likens our relationship with the earth to that between a mother and a child. It's not just that we love nature; "the land loves us back," she wrote in *Braiding Sweetgrass*. "She loves us with beans and tomatoes, with roasting ears and blackberries and birdsongs. By a shower of gifts and a heavy rain of lessons. She provides for us and teaches us to provide for ourselves. That's what good mothers do."[8]

As with any long-term relationship, committing ourselves to being in our world starts with love. I found love in my backyard garden and the endless opportunities to meet the natural world through my senses: kneeling on the earth to tend to the seedlings, stretching to pull a weed, hearing the birds singing all around me as I worked. I smelled the earthiness of the soil and tasted the life in just-picked snow peas and lettuces.

These experiences are available to all of us, garden or not, when we fall in love with the world around us. We don't have to manufacture love; it emerges spontaneously through *being* together: saying yes to reality and practicing mindful presence. Love then inspires us to *think* and *act* in ways that strengthen and sustain our relationship, forming the self-sustaining cycle of Think Act Be.

Think: Understand the Pull

For most of us, focusing our attention on a screen has become second nature and we rarely question the forces that drive our patterns of use. When we understand the pervasive pull of technology, we can find more ways to avoid its unwelcome intrusions.

RECOGNIZE REWARDS

The biggest challenge with addictive apps is the immediate reward they offer—or, more precisely, *might* offer. The tools that meet our needs every time, like our calculator or calendar, aren't the ones that call to us when we're trying to live our lives. We're not wasting hours on the weather app, compulsively calling people, or scrolling through our bank transactions yet again.

The apps that pull us in give us something we crave—a laugh, entertaining or outrageous news stories, approval from others. But the trick is that they give it to us only *some of the time*. When I watch comedy clips on YouTube, a few are hilarious but most are just okay.

Sometimes we get the likes and shares we're hoping for on social media, and sometimes we don't. Sometimes the news is juicy, but it can also be boring.

In behavioral terms, we're on a "variable ratio reward schedule." That means a particular action leads to a reward only some of the time. A slot machine is the archetype of variable ratio reward in humans; you never know when you're going to win, and the next pull of the lever could be the jackpot. The setup is the same when we open an app, click a video link, or read a post; we have to repeat the behavior some unknown number of times to get the reward.

Maybe the first Instagram post we read is really inspiring, which attaches an emotional tag to opening Instagram. But then we have to scroll through thirty more to find another one that speaks to us and activates that emotional tag. Sometimes the email we're expecting is waiting in our inbox the first time we check; more often, we're disappointed repeatedly before it arrives.

Notice what pulls you into your phone. What is the reward that your brain might expect? Does the reward happen consistently, or is it variable? See whether there's also any fear of missing out on something by not opening certain apps.

It's really hard to stop a behavior when our brains know that eventually a reward will come, even if the behavior is sucking the life out of us. That was often my experience with my smartphone. When I just needed to unwind, it still called to me. "Come on!" it said. "We've had such fun together! I know you looked at your email sixty seconds ago . . . but just check it again. And check the headlines while you're at it. Maybe there's something special there for you."

I wanted to say no but rarely did. Little did I realize that it wasn't my rational brain that kept me glued to my phone. The crucial brain regions for the emotional reward I sought were subcortical—tucked away deep in the brain, below the level of conscious, deliberate thought.

The addictive power of technology depends on the same parts of the brain[9] that addictive drugs such as cocaine act on, along with food, sex, shopping, and gambling. It's all reward, encoded through the brain's dopamine system. Cocaine-addicted rats on a variable ratio reward schedule will press a lever like mad until they get a dose injected into their brains—sometimes after five presses, sometimes two, sometimes twenty. The next press could always be the one that delivers the hit they're after.

My self-defeating phone use was driven by my unwittingly seeking the emotional tags that had rewarded my past phone use. Often, the reward we find is relief from anxiety. In the movie *Adaptation*, screenwriter Charlie Kaufman (played by Nicolas Cage) sits at his typewriter, staring at a blank page. His anxiety is palpable as he struggles for words, and his voiced-over thoughts turn to his coping mechanism—food. He considers getting coffee, or maybe coffee and a muffin. He tries to write again, but his anxiety keeps pushing him toward escape.

Just as Kaufman's anxiety pushed him toward distractions, our anxiety drives more of our compulsive screen checking than we may realize. A review of anxiety and smartphone use suggests that reaching for our phones is often "an experiential avoidance strategy to deflect aversive emotional consequences."[10] In other words, anxiety is uncomfortable, and our phones offer innumerable distractions to relieve our distress. However, the relief is short-lived, and the distraction must be repeated again and again. This form of escape also prevents us from developing more effective ways of coping with anxiety.

In hindsight, I felt as if I had become a tool that my smartphone was using instead of the other way around. My attention wasn't my own but was being siphoned off into my phone. Although it was nice to have immediate relief from boredom, I found that my free time was no longer free. Often, when I just wanted to relax at the end of the day, I would instinctively pull out my phone, worried I might miss the enjoyment I could have on it. Surely I could find something to entertain me if I looked through enough apps. I'd often wind up reading news stories that didn't interest me and only made me feel worse, both mentally and physically.

One day it hit me that no matter what I did on my phone, the scene never really changed. I was looking at a screen that offered a two-dimensional facsimile of the world and that generally delivered less than it promised. While my mind was being continually drawn into that digital world, my body and spirit knew that something essential was missing. They were pushing me toward the full three-dimensional world. I needed to engage my other senses—to hear the sounds around me, feel myself in contact with matter, read an actual book. And yet despite feeling fed up with my phone, I found myself robotically reaching for it, often drawn in by the stories my mind was telling me.

IDENTIFY THE STORIES

Like many of the people I've treated, Marc wanted to cut down on his screen time. Marc's work often kept him tied to his phone, and once he was on it he'd check his personal email, social media, and the news. He wasn't buried on it for hours at a time, but it was always close at hand, and every hour was punctuated by several interactions with it.

Marc had tried many times to limit his phone use, but he found that the siren call was irresistible as long as his phone was available. He wanted especially to be more present for his two-year-old son, Max, and also for his wife, who was unhappy about his constant phone use.

The changes Marc made started with recognizing the mental stories that kept him tied to his phone. He believed that bad things would happen if he left it at home for an hour while he took Max to the park: "My boss will be mad she can't reach me." "I'll miss a client's call or an important email." Perhaps these stories were true, but he began to consider alternative outcomes. In reality, he almost never got urgent calls at the park, and even when he missed an important call while he was in a work meeting, it was not a problem to return the call when he was available. On a deeper level, Marc realized he was willing to take some minor risks if it meant more connection with Max.

Marc also saw through the belief that he could take his phone with him but leave it in his pocket. He realized that a more accurate story was "If I take it, I'll check it." Addressing these thoughts and beliefs was an essential part of what helped Marc to find more balance and to change his behavior.

Changing our relationship with technology can be really hard—even when we feel the costs of too much screen time and long for more connection to the world and the people in it. Some of the most effective changes we can make are behavioral; our heart and head might say yes to staying present, but it's our hands that steer the ship.

Act: Bind Yourself to the Mast

The idea of separation from his smartphone sparked more anxiety than Marc had expected as his mind crafted stories about bad things that would happen. Being without his phone would be a kind of exposure therapy as he deliberately faced his fear for the sake of something more important to him. Marc discovered the answer to his mind's fearful protests when he left his phone at home on family walks and on trips with Max to the playground.

Through these behavioral experiments, Marc realized it was okay to be unavailable sometimes and that pretty much anything could wait for an hour or so. The initial unease he felt the first time he went phoneless was mixed with excitement. How freeing it was to have nothing pulling his attention away every few minutes. He had made a single decision that eliminated countless future decisions about whether or not to check his phone. As a result of his greater presence, Marc found that it was much easier to connect with Max and with his wife.

Cognitive insights led Marc to *act* and unchain himself from his phone. With his attention no longer pulled into various apps, he experienced greater awareness and *being*.

CONTROL AVAILABILITY

As I started my self-help CBT, I knew my own resolution to spend less time on my phone would be ineffective as long as my phone and the apps on it were readily available. In the past I'd tried heart-and-head approaches, such as when I resolved to put my phone away during our family's week of summer vacation. I had planned to check it in the mornings before anyone else was up, and that was it. *No exceptions!* Except that *wasn't* it because I would use it for other things, such as navigating to the zoo or streaming music while I cooked dinner. Every interaction with my phone was a chance to bend my self-imposed rules ("I'll just check my email . . .").

The next year, I decided to leave my phone and computer at home for our week at the beach. I thought I'd miss them, but it turned out to be quite the contrary. I'd forgotten what it was like to have nothing between me and the world. When we got home, I removed all the apps that I knew I would use compulsively. There are times when it would be more convenient to have easy access to those apps on my phone, but the benefits have outweighed the costs. The vast majority of the time, I just feel freedom from not having that option.

Each of us has different needs and wants, and we will draw our own idiosyncratic boundaries around technology. None of these is completely foolproof, and as I've found over and over, we're good at finding workarounds. But simply creating a little separation between ourselves and our technology gratification can give us more choice in how we spend our time.

The most effective strategies for changing our habits are often action oriented, as captured by Homer in his epic poem *The Odyssey* (resisting temptation is apparently an ancient problem). The title character, Odysseus, wanted to hear the song of the Sirens, but he knew he wouldn't be able to resist their call and would steer his ship into the rocks. So he had his men plug their own ears and then tie him to the mast of the ship before they reached the Sirens. That way, he could hear their song but wouldn't have the option to act on his overwhelming urge to navigate toward them.

This is such a powerful principle. Our motivation to change a behavior *tends to be lowest when we need it the most.* Odysseus had strong motivation not to wreck his ship when he was far from the Sirens, but he knew he'd be powerless to resist them when he heard their singing. We can resolve to drink less, or forgo dessert, or get up at 5:30 a.m. to exercise when we're not actually faced with the option to have that last glass of wine, or eat a pint of butter pecan ice cream, or reset our alarm for 7:00.

Consider a change you might want to make in your relationship with technology. What would binding yourself to the mast look like? Good options include anything that will make the new behavior easier and/or the old behavior harder.

By making commitments that are hard to take back, we can import our high motivation into times when we know our motivation will be lower. I know that when I get an urge to check email on my phone, I'm going to ignore the little voice that says, "Do you really need to do that right now?" But I've bound myself to the mast by leveraging the *act* principle of making it harder to do what I don't want to do. I'm not going to go through the trouble of reinstalling an app every time I have an urge to use it. It's not worth the time and effort.

LIMIT NOTIFICATIONS

Discrete mealtimes limit the sprawl of food into every moment of our lives. Imagine how much more we would eat if our kitchens were always sending us notifications: "Are you hungry now?" "Want some chips?" "How about an apple?" "It's been ten minutes. Maybe some mixed nuts?" Many people find it helpful to turn off nonessential notifications for a similar reason—so their phones aren't constantly beckoning.

FIND EASE

Changing our relationship with technology doesn't have to be a white-knuckle struggle. We can let it be easy with the leverage that Think Act Be provides. By being in the present and saying yes to our experience, we fall in love with life unfiltered. We simply no longer want a screen between us and what we love. With our thinking minds, we can better understand our relationship with technology and make plans for abiding in the real world. Through our actions, we make effective commitments to experience more being.

Reducing overexposure to technology is never out of reach. It's as near as putting down a screen and looking up and realizing there's a whole world out there—your world and my world, the full three dimensions that surround us, the sky above us and the ground below. This ultra-high-definition world in infinite shades of color is always present, extending in every direction as far as we can see. By being in it, we connect with ourselves in all our dimensions and clear the path toward mindful awareness.

Making ourselves available for more of life doesn't always lead to profound experiences of joy or gratitude, any more than planting a garden in an unshaded spot guarantees the sun will shine. If we set up expectations for feeling peace or transcendence when we put away our screens, we're bound to be disappointed. Sometimes things are boring, or disappointing, or irritating.

But endless possibilities open to us when we make ourselves available for more of life. Subtle changes in our behavior can lead to profound shifts in our lives, even if the changes seem inconsequential—what's the big deal about reading the news while in the emergency room, or taking our phones to the park? But, as we've seen, these everyday choices are like micro-adjustments to the steering wheel; over time they keep us on the road. Anything we do will be enhanced by bringing greater awareness to it. Our heads will thank us for the mental break. Our hands will relish the opportunity to be of use. Our hearts will enjoy the resonance with reality, like feeling the full warmth of the sun.

There's no room in this approach for feeling deprived. On the contrary, the overriding response to this kind of engagement is gratefulness. Being filled with thankfulness in ordinary or even challenging situations is one of the first fruits of mindfulness. In the next chapter, we'll see how the practices of Think Act Be foster gratitude.

OFFER THANKS

You probably know that being grateful is good for you, but it's often hard to connect with gratitude. In this chapter, you'll learn that you don't have to try to make yourself feel grateful; instead, gratitude emerges when you open your awareness to the gifts all around you. You'll also learn about the thought patterns that preclude gratitude, as well as specific actions that can train your awareness to notice good things in your life. Gratitude in turn makes us more willing to accept our lives as they are, which supports our mindful awareness.

* * *

I have very clear memories of singing the exuberant hymn "Count Your Blessings" countless times in the churches of my youth. Even as a kid, this singsong call to gratitude struck me as an unrealistically sunny response to hardship. I doubted that the songwriter had any idea how difficult his exhortations were.

Are you ever burdened with a load of care?
Does the cross seem heavy you are called to bear?
Count your many blessings, ev'ry doubt will fly,
And you will be singing as the days go by.

And yet, from what I've gathered about the prolific hymnist Johnson Oatman Jr., he may have been writing for himself. Oatman is said to have been in "failing health" in 1893, four years before he wrote this song.[1] I find it reassuring to know that he wasn't cheering from the sidelines of suffering; he knew his own share of pain. Maybe this wasn't a shallow invitation to gratitude but a stance of defiance: *circumstances be damned, I'm going to give thanks.*

There's no denial of pain here. We're "burdened with a load of care." Our cross is heavy. As another verse says, we're "discouraged, thinking all is lost." And we resolve to see that our pain isn't the whole story and doesn't get the last word. This defiant gratitude is one of the nicest things we can do for ourselves—especially when we're down and disheartened and *least* feel like giving thanks.[2] The countless benefits of gratefulness include greater life satisfaction,[3] improved mood,[4] and stronger relationships.[5] But how can we find gratitude when our lives feel far from all right and we're more likely to feel bitter than grateful?

I had my share of bitterness during the worst part of my illness as I kept thinking, "This shouldn't be happening." I was frustrated that I couldn't get back to the way life was supposed to be. On a particularly painful night, I received a message from the contemporary Stoic philosopher William Ferraiolo about his latest book on Stoicism. A passage from the book was exactly what I needed. "You have suffered, you will suffer much more, and a lifetime of your suffering will culminate in your death," it read. "When you can muster genuine gratitude for all of that, then you will have made the kind of progress that is not easily reversed."[6]

They were hard words, and yet I found great comfort in them. My struggles were not a strange aberration, beyond the bounds of ordinary life. To live is to face suffering. In that moment, pain was my reality. I could choose to receive it rather than fighting it. I could step into mindful acceptance.

There are tremendous benefits to being grateful, but many of us see gratitude as a guilt-filled obligation—as the *nice* thing to do, the *good* thing. Gratitude can start to feel like another moralistic thing that we're "supposed to do," like eating right and flossing our teeth. Gratitude is what you give in exchange for a gift. If you don't respond with gratefulness, you've left the store without paying for your purchase. Gratitude completes the transaction.

But treating gratitude as an obligation defeats the purpose, like forcing a child to mumble a very ungrateful-sounding "Thank you." We don't have to manufacture feelings of gratitude, as if trying to squeeze one more glob of toothpaste from an empty tube. It's available no matter what's happening in our lives, and we can foster it through the simple practices of Think Act Be. Being thankful starts with paying attention.

Be: Shift Attention

You may have seen a humorous Christmas video[7] that came out a few years ago. A man is lying in bed in the opening scene, covered in wrapping paper. As he unwraps his head he exclaims, "I'm alive!," as if it's the best present he's ever opened. The video follows him through his morning as he turns on the lights—"Honey, the power works!"—and discovers that they have clean water from the tap. The gifts continue: a shower, shoes, breakfast, a briefcase, coffee . . . even a car! There's no sense of obligation in the man's grateful excitement, just pure delight.

The video is funny because it's so different from the way we typically greet the day. We often miss out on gratefulness because we're programmed by evolution to notice what's wrong in our lives, like reporters looking for bad news. Noticing problems is an invaluable ability and a necessary step toward solving them. We need to know when we're sick, or if there's a problem at work or in a relationship. But the fraction of our lives that isn't perfect routinely crowds out all the good, like filling the evening news with one distressing story after another. I can walk outside on a gorgeous day in my safe neighborhood but ignore the good things around me and see only the problem on my mind. It's easy to have the impression that nothing in my world is going right.

This impression is bolstered by our habitual tendency to hit "Tare" on the scale of life, zeroing out every good thing we have. But just waking up in the morning is far from nothing—we have another day! And we're lying in a warm bed, not in the street. We're in a house. We have a body. Gravity glues us to the floor when we stand. We pad to the bathroom, which is *inside our house*! Food awaits us in the kitchen. A true tare would start with an empty scale and would weigh the full measure of these gifts.

When I was a kid, my brothers and I got the "gratitude talk" every few months. We would be living our kid lives, thinking things were going pretty well, until we heard our dad say, "Boys, I need to talk with you about something." *Oh, shoot.* The talk usually began with "There hasn't been a lot of gratitude around here lately." "Rats," I would think. "I meant to remember to be grateful after the last lecture!"

I felt a lot of guilt at the time, but the trouble I had with being grateful wasn't a moral failure. I had just stopped paying attention. I wasn't seeing everything that my parents did for me and that life freely offered. It took me a long time to realize that gratitude isn't an emotion we have to drum up. Like anything we offer back in this life, it's a gift we give ourselves (so be sure to thank yourself!), and it starts with paying attention.

> **Try this simple breath-centered gratitude meditation. With each breath, bring to mind one good thing you have. Inhale, exhale, "a warm house." Inhale, exhale, "raisins on my cereal." You can include things you experience every day, like a refrigerator or warm socks, or singular events like running into a friend at the grocery store.[8] You might use this practice if you're struggling to fall asleep. Rather than bemoaning being awake, take the opportunity to remember the good, like being awake in a warm bed, or not living in a war zone. (Just be careful not to turn it into a tool to try to make yourself fall asleep, which will often backfire.)**

By practicing mindful awareness, we can expand our attention beyond the subset of our lives that we consider problems and notice everything that's working. Rather than being overwhelmed by what's wrong, we can be overwhelmed by all that's right. This type of attention is especially useful when we're focused on everyone who seems more fortunate than we are.

CHANGE YOUR COMPARISONS

After being ill for a long time, I wondered bitterly why I was sick when everyone around me seemed to be healthy. "Look at all those active dads out playing with their kids," I complained to myself. Focusing on their apparent health and vitality amplified my sense of not getting what I deserved, and I started to think I was getting a real bum deal.

We often fall into these upward comparisons as we dwell on the relative fortune of others, which further fixes our attention on our problems and deprivations. But our daily hassles and even our bigger problems can fade into insignificance when we remember how much worse things could be.

We rarely have to look very far to find those who would gladly trade their situation for ours. I came to see how I was ignoring all that was still going well even at the lowest points of my illness and the ways my life could have been immeasurably worse. Many times, I was pulled out of self-pity by hearing stories about others' misfortunes, like a stroke that had taken away a person's ability to walk or speak, or parents who had lost a child.

This is not to deny our own struggles, and even less to forbid others from complaining about theirs. Most of us don't appreciate being told, "Be grateful—things could be a lot worse." But shifting our comparisons can help us keep our troubles in perspective. No matter what we're going through, there are countless potential calamities that we are *not* facing in any given moment.

We can even do downward comparisons with ourselves, remembering times when things were worse. Yesterday, I dealt with nausea for most of the day. When I suddenly noticed midway through today that I wasn't the least bit nauseated, I put both fists in the air with a triumphant "*Yes!*," as if I'd just hit a walk-off home run.

Objectively, I felt the same physically as I had a moment before, but the mental contrast with being nauseated made my current state feel like a huge win.

We've all experienced these contrast effects before—we recover from a stomach bug and feeling normal feels amazing. We come inside from shivering in the cold and our warm house feels heavenly. We welcome the sounds of our kids arguing after a strangely silent week when one of them was hit hard by a virus. Our mental machinery quickly habituates to our circumstances and notices relative changes much more easily than the status quo.

Take a few moments to consider what's going right for you right now. For example, are most parts of your body working well? Do you know where you'll find your next meal? You can also consider all the things that *could* be going wrong but aren't right now, like having a massive cold or being stuck in a fight with your partner. Don't try to force any feelings; just direct your attention with an open mind.[9]

We don't have to wait to recover from a sickness or come in from the cold to experience a favorable contrast. Most of the time when we're feeling neutral to slightly negative, we can discover a bit of gratitude by recalling how much worse things have been. When I remember the worst of my sickness, I shake my head in amazement at how much better I feel most days now.

FIND THE GOOD

When my wife and I were in France for our junior year of college, we had to carry our groceries back to our dorm, since we didn't have a car. I often felt annoyed when the handles of the heavy plastic bags cut into my hands as we walked. But although the pain was unpleasant, it also pointed to something very good: we had plenty of food.

A path to gratitude is often found right in our everyday disappointments or frustrations. It's annoying when our flight to a fun destination is delayed—and I can realize what a privilege it is to go on vacation. I don't like it when our dishwasher breaks and needs to be replaced—and thank God we can afford a new one. Having to stop for gas is irritating when I'm in a hurry—and I have money for gas and can even pay at the pump. A family-wide norovirus outbreak is truly miserable—and it underscores that I'm not alone.

Years ago, a patient of mine had a dramatic experience of this type of gratitude. A few months after the birth of her second daughter, Julie was taken to the hospital with shortness of breath. The doctors discovered that her pregnancy had provoked a serious heart condition that required immediate surgery to save her life. Julie told me that her relationship with life changed profoundly after the successful operation. She was grateful for everything, even the painful follow-up medical procedures during her recovery. To hurt meant she was still alive and had more time to spend with her husband and two little girls. Every morning when she woke up, she reminded herself that it was a gift to open her eyes.

Think of a recent challenge you've gone through. Without denying your struggle, was the difficulty pointing to anything positive? For example, stressful work highlights that you have a job to pay your bills; going to the emergency room means that

you have access to twenty-four-hour care. Be on the lookout for other difficulties that reveal the good in your life.

The Benedictine monk David Steindl-Rast pointed out that we can be thankful for the chance to address problems in our life—our child's sickness, environmental devastation, a leaking roof, a broken relationship. "We cannot be glad for those things in and of themselves," he wrote, "but we can be thankful for the opportunity to do something about them."[10]

As with downward comparisons, beware of using this practice to invalidate the difficulty or discomfort that you or someone else is experiencing. Mindful awareness doesn't push anything away; rather, it expands the bounds of our attention to change our relationship with our problems. Gratitude in the face of suffering is about not denial but transcendence.

As we practice expanding our attention and noticing more that's right, we can receive even our challenges with gratitude for what they teach us and how they require us to grow.

DISCOVER GIFTS

These days, when I find myself crying in the shower or on the couch, it's more often with gratitude than from feeling sorry for myself. I am so relieved to be in better mental and physical health than I was at my lowest. Nevertheless, I can't deny that my extended illness was a gift in many ways. It took a long time to recognize these gifts—for years, I saw my health problems as nothing but a curse that I needed to undo. But eventually, I came to see the good they had brought into my life.

My struggles gave me deeper compassion for others and helped me understand their suffering. Some of my most meaningful work as a therapist emerged from those dark times. I became more willing to accept my limits and to let others see my weaknesses. I discovered deeper love from my wife and kids, whose life-giving devotion during my sickness still makes me cry. And if I hadn't reached the end of myself (as I described in chapter 1), I wouldn't know what's on the other side.

It's often hard to find gifts in our suffering. I'm still good at self-pity at times, and despair can find me on difficult days. But if given the option, I wouldn't trade my sickness for all it taught me. I've even worried at times that if I fully recovered, I would abandon the spiritual truths I had known, like the man who promises to devote his life to God if he can just find a parking space for a big meeting and then, when a space opens up, says, "Never mind, I found one!"

I'm not alone in finding gratitude for the growth that comes through difficulties. Many of my patients have had similar realizations as they emerged from a difficult struggle. Antonis, for example, described a profound sense of accepting the painful depression he had come through—of saying yes to that experience—because it shaped him into the person he was. Few things are more empowering than seeing how our pain transformed us into a new creation that could only have been built from the broken pieces of our life.

This quality of awareness and *being* can lead to changes in how we *think*, allowing us to rewrite some of our fundamental assumptions about life and what it owes us.

Think: Examine Beliefs

There's an easy way to predict how much gratitude we'll experience. We simply add up all the good things we have in our life and then subtract the things we assume we're entitled to. What's left will correspond very closely to our level of gratefulness.

CHALLENGE EXPECTATIONS

We often don't recognize our assumptions about what we deserve because they reflect our unconscious core beliefs about the world: Life should be painless. We shouldn't get sick. Our physical comfort should be uninterrupted. People should be nice. These assumptions set us up to feel cheated and disgruntled by inevitable disappointments. If life doesn't meet our expectations, then surely there's been a mistake.

When I was a kid, my dad would often respond to our bellyaching with lines like "You want fair? Fair is a bowl of rice a day." I didn't get it at the time. I kept assuming that the world should conform to my expectations for what was coming to me. But one of the most freeing realizations I've had is that *life owes me nothing*. Not the average life expectancy. Not physical comfort or stress-free days. Not good health. Not praise and appreciation for all I do.

I was stunned when I realized that this was the deal. "But surely life owes me . . . *something*—right?" I asked myself. A minimum number of years? Living to see my kids reach adulthood? Avoiding certain kinds of pain? But no. Seriously, buddy—nothing. Not even a single problem-free day.

I had to relearn this lesson during my illness. In my mind, there was only one acceptable answer to my prayers for relief: the removal of my suffering. I interpreted my continued struggles accordingly as an inconsiderate nonresponse. It was many months before I recognized the gifts that had carried me through my sickness and depression. I was surrounded by loving support from my family, and I was granted strength again and again to face each challenge.

It can be paradoxically comforting to realize that our unwelcome problems aren't caused by a glitch in the universe. Anything we have—including life itself—is a gift. "Life is given to us; every moment is given," wrote David Steindl-Rast in *Music of Silence*. It can seem so obvious, but most of us forget it most of the time. "The only appropriate response," continued Steindl-Rast, "is gratefulness."[11]

QUESTION THE STORY

One of the most common mental stories that blocks gratitude is that "this shouldn't be happening to me," as if life's burdens are misplaced when they land on our own backs. *Sickness?* Not mine. *Inconvenience?* Definitely not mine. *Disappointment?* Surely there's been a mistake. Carrying hardship that we don't think is ours feels like picking up someone else's dirty clothes from the floor. *This should not be my problem!* we protest.

For years, I was sure my illness had been sent to the wrong address. I had looked after my health, eaten well, exercised, spent time with friends and family. It didn't make sense that I had become this tired, silent, sullen person. But my insistence that this shouldn't be happening to me just made it worse. My burden felt a lot lighter when I accepted that it was mine to carry.

We can use cognitive techniques to examine thoughts that crowd out gratitude, like the black-or-white thought that "nothing was going my way" during the course of my sickness. I was narrowly focused on what *wasn't* right and ignoring all that was, like a visitor to the Louvre who notices only the toilet that's clogged. When I asked myself whether my belief was definitely right, I realized that nothing could be further from the truth. *I could breathe!* That wasn't nothing. I could stand. I could walk, even through the fatigue. My voice impairment made it hard to verbally express my love to my kids, but I could still hug them.

Simply focusing on the breath, without bringing anything else to mind, can itself be a gratitude practice. Your breath, after all, is what's keeping you alive. Hold this realization in wordless awareness as you offer thanks for this breath . . . and this breath . . . and this breath . . .

There are many ways to remind ourselves of what we have, from the air that fills our lungs to the people who fill our lives. With practice, we can train our minds to recognize the gift of life in all its manifestations, beyond the narrow sliver of our expectations. We can choose to direct our attention to the 95 percent of our life that is good instead of the 5 percent that's imperfect.

Act: Practice Gratitude

Although gratitude can't be forced, Think Act Be offers many practices that open the door for this experience.

JOURNAL

My overarching sense at the end of each day during my depression was that *that was rough.* My attention predictably was drawn to negatives rather than positives. But even on days that felt like pure struggle, there were still many good things that had happened. I had a hot shower. I enjoyed good food at every meal. My family still loved me.

Writing down the things that went well at the end of each day is a very effective way to direct attention toward the good in our lives. Many studies have shown that this simple practice can evoke gratitude.[12] Journaling can also help us be more attentive during the day to what's going right—the antidote to the belief that "nothing is going my way."

As part of my own cognitive behavioral therapy, I starting keeping a journal next to my bed and writing down at least three good things from the day right before I went to sleep at night. If you try this practice, it's better to be specific rather than general; for example, instead of writing "my kids," I might write "playing catch with my kids after dinner." Be careful not to fall into rote repetition, like "Air. Water. Food. Boom, good night."

I recommend getting a new journal for the practice. I had done it in the past on scraps of paper or sticky notes, but it felt more appropriate to write in something dedicated for the purpose. Watching the pages fill up can be an impressive reminder of all the good in life. As with any gratitude practice, just focus on the things you're listing rather than trying to conjure a feeling of gratitude. Evaluating whether it's "working" is likely to take you out of the experience.

SAY "THANK YOU"

One of the most effective ways to practice gratitude is also one of the most obvious: just say "thank you" to someone. Being on the lookout for opportunities to express our gratitude primes us to notice good things we could otherwise miss. Research shows that even writing about the people we're thankful for boosts our well-being, but the greatest benefit comes from actually expressing our gratitude to the person.[13]

We can practice at home, where the people we live with tend to get "tared" with the rest of our daily experiences; being noticed and appreciated is extremely beneficial in our close relationships. We can also bring the practice to people outside of our home and offer thanks for services that are often taken for granted, like the fishmonger's care in selecting our seafood and our mail carrier's rain-or-shine service.

In general, the more specific the appreciation, the better. Blanket statements like "Thank you for all you do" can sound bland and perfunctory. We'd usually prefer to be told something like "I love how much care you put into arranging the berries on the tart" versus "Thanks for making dessert." Being precise shows a greater level of attention to the other person's actions.

Gratitude rests not on the specifics of our situation but on our internal willingness to find joy. No matter where we find ourselves—rich or poor, in sickness or in health—we can find gratefulness simply by vowing to stay with our experience.

OPEN TO LIFE

There are many ways to invite gratitude, and they all share a common element: being present. At an elemental level, the willingness to open to our lives is itself the essence of gratitude. With openness, we say yes to whatever life offers, from the brightest moments to the darkest. We receive creation exactly as it is.

That essential willingness is a spiritual practice—an act of worship. Opening to all of reality connects us with the constancy of our spirit, the unchanging, ever-present witness to experience. Our spirit isn't locked into up-or-down judgments of our situation: good/bad, for me/against me, love it/hate it. This spiritual attunement fosters gratitude, which in turn nourishes our spirits.

Gratitude is implicit in our willingness to abide in the real world. It was gratitude I discovered that night in the emergency room with my daughter when I resisted my habitual tendency to zone out on my phone. The "nothing" experience of waiting in that room as Ada slept was transformed into an everything experience, just by my being aware that I was there and that, mercifully, all was well.

Gratitude gives rise to more gratitude as we recognize it as a gift. We are grateful for feeling grateful. "When we are thankful for whatever is given to us, no matter how difficult," wrote David Steindl-Rast, "no matter how uninvited it may be, the thankfulness itself makes us happy."[14] From this perspective, there can be no guilt about not feeling grateful, since we're withholding primarily from ourselves. Thanksgiving is counted among life's innumerable blessings.

There is a contentment that's deeper than our circumstances. As the Apostle Paul wrote to the Philippians, "I have learned the secret of being content in any and every situation, whether well fed or hungry, whether living in plenty or in want."[15] Gratitude is the key. "Give thanks in all circumstances," Paul said elsewhere.[16] James went even further: "Consider it pure joy, my brothers and sisters," he wrote, "whenever you face trials of many kinds, because you know that the testing of your faith produces perseverance."[17]

Maybe we'll have moments or days when we say "To hell with gratitude—this is too hard." Sometimes things are just hard and we can't thank our way out of it. Sometimes we want to cry at the unfairness of it all—and we can be in *that* experience, too. That's where we are. Being grateful doesn't mean you never feel sad for yourself and never wish for relief from your suffering; gratitude and self-compassion are perfectly compatible. We can't shortcut our way around our trials, using gratitude as a "get out of pain free" card. Sometimes we have to get lost in our suffering before we can be found.

Perhaps part of the gratitude we find comes from knowing that the universe makes space even for our frustration and self-pitying tantrums. It welcomes our selfish prayers and our bitter honesty—probably more than our trite gratitude through gritted teeth.

I certainly have those days when gratitude feels far away. I know being thankful is an option, but I choose instead to grumble and sigh about what's going wrong. I'm learning not to feel guilty for these episodes. There's even strong precedent for them in sacred texts. The same book of the Bible that proclaims "I will sing of the Lord's great love forever"[18] also implores, "How long, O Lord? Will you forget me forever? How long will you hide your face from me?"[19]

So, if you're feeling "burdened with a load of care,"[20] as Johnson Oatman Jr. wrote, don't worry about trying to make yourself feel grateful. Start by receiving the load as your own. Gratitude begins when we say yes. Abide in the experience. Simply notice whatever you have: the pain, the struggle, the means to respond and to grow. When we willingly enter into every situation we encounter, we can discover joy and contentment that transcend our circumstances.

Gratefulness that's grounded in mindful *being* will fundamentally alter the way we *think* about life. It will guide us to *act* in ways that foster more mindfulness and gratitude. And it's closely tied to finding the rest we need for our bodies and minds, as we'll see next.

FIND REST

Rest is foundational for coming home to ourselves and finding peace and connection. In this chapter, you'll learn why mindful trust is essential for rest and how mindfulness offers you greater ease even as you're working. You'll also discover the types of thoughts that often make it hard to find true rest and how to counteract them. As you'll see, you can rest even from wanting life to be different from the way it is, which opens the door to ever greater mindful acceptance and connection with your experience.

* * *

In the first few years of my private practice, I paid lip service to the importance of self-care and stress management, always with an asterisk in mind (*time permitting). It was an intense time in my life, with three young kids, a full-time teaching position at a local college, and a growing private practice. Even as I helped my patients deal with overwhelming stress, I rarely allowed myself the time I needed to manage my own.

When my regular hours filled up, I started seeing patients on Saturdays and by videoconference from home late at night. I resisted taking time off, even when I was sick. I thought I might be working too much, but I found it almost impossible to reduce my hours. I couldn't stand the thought of turning away people who needed the kind of treatment I offered. I also feared for my family's financial security, since my temporary teaching position was going to end soon. "What if I don't have enough hours in my clinical practice?" I worried. "How would we afford our mortgage and health insurance?"

Soon enough, the long hours and constant stress were catching up with me. I was more irritable at home, to the point that my then-five-year-old would often call me "Mad Dad." Family commitments became unwelcome intrusions, taking time from my work that I didn't think I could spare. I started to drink more to unwind at the end of every day. I no longer had the enthusiasm that had made me want to be a therapist years before, and the relief I felt when the weekend came was overshadowed by the dread of knowing I'd be back at work on Monday. Even a week's vacation felt much too short. And yet I persisted. "I should be able to do this," I silently insisted. "I just need to get better at managing stress." But my "should" didn't change the reality of my burnout.

It took a brush with death to open my eyes to the heavy toll of unmanaged stress. One evening, as I biked through an intersection on the way home from my office, I looked to my right and saw that a car was going to run a full-on red light. I slammed on my brakes just in time as the car sped past me. I felt lightheaded as I realized I could have been killed. A second later I thought, "At least all of *this* would be over." That startling thought made me realize that a part of me would welcome death as the end of all the strain.

How can we reduce stress and find effective ways to rest in the midst of our busy lives? That question guided the major changes I made in the coming years, and it ultimately helped me discover the full measure of rest that's available to all of us.

Recognize Stress

When I saw from the corner of my eye that that car wasn't going to stop, my sympathetic nervous system prepared my body for action so I could avoid the danger. My adrenal glands pumped adrenaline into my bloodstream, my heartbeat quickened, my breathing sped up, and many other changes helped me avoid the grille of the car.

Moments later, my body also turned on the hypothalamic-pituitary-adrenal (HPA) axis, which causes the adrenal glands to release a different set of hormones, such as cortisol, which keeps the body in an activated state. You're probably familiar with the feeling of elevated cortisol during periods of high stress, like final exam week or a difficult stretch at work. It's really hard to relax when we're vibrating at this higher frequency.

As I pedaled home, I could feel the lingering effects of the adrenaline as my heart pounded and my legs felt wobbly. I was more vigilant than usual, taking extra care at intersections to ensure that the way was clear before continuing. By the time I was home, my parasympathetic nervous system had taken over with the so-called rest-and-digest response that counteracts the effects of the sympathetic system. The extra tension had drained from my body and mind, and I felt relatively calm again.

Our nervous systems are exquisitely designed to handle short bouts of stress. In the face of a threat, our brain triggers a full-body response to keep us safe, preparing us to fight an enemy, take flight, or freeze in place. Having a robust stress reaction in life-threatening situations is a great thing; it turns on to protect us from an acute threat and quiets down once the danger has passed.

But for many of us, the stress we feel isn't a rare and fleeting experience; instead, we feel tense all the time, as if we were living in a never-ending final exam week. The dangers we fear often aren't even real; instead, we feel the stress of scary things we *imagine*, like missing our train or losing our job. Rushing busily from one commitment to the next compounds our stress as it fosters an unrelenting sense of urgency. This nonstop pressure and perception of threat keep activating the sympathetic nervous system and HPA axis. Eventually, the constant flood of stress hormones takes a major toll on our minds and bodies—even as our crowded lives feel strangely empty.

Anything that demands a response from you can be a source of stress. What are the major sources of stress in your own life? Consider how the stress is affecting your body and mind and maybe even your health. Do you need more effective ways to manage stress?

The ripples from my own unmanaged stress continued to affect me for years and contributed to my eventual chronic illness. The disrupted sleep I experienced is often an early sign of being flooded with stress. Other indications include physical tension, anxiety, irritability, digestive problems, alcohol abuse, and depression—all things I experienced as the stress continued. Physical illness often follows, as it did for me.

Nevertheless, it took me a long time to recognize the role that stress played in my physical health. At the peak of my sickness, I often felt that my whole body was buzzing, like an alarm someone had forgotten to turn off. I would find myself constantly lifting my shoulders in a half shrug as if I were bracing for impact. This physical tension fed into a background sense of unease. I was always on edge and startled by everything, and my kids' joyful shrieks would feel like slaps to my nervous system.

When I realized how stressed I was, I turned to the usual suspects to take the edge off—monthly massages, vacations with my family, daily meditation—but the relief was short-lived. Sometimes these activities even fed into my stress. During a week at the beach, for example, I constantly had my finger on the pulse of my stress level: "Am I relaxed now? What about now?" Each time I checked in with myself, I felt a small spike of anxiety, since vacation felt like my high-stakes "one chance to relax" before returning to the unrelenting pressure that was waiting for me. I had traded stress about work for worry about stress.

Rest from stress can elude us even when our bodies, minds, and spirits desperately need it. Common techniques for managing stress, such as relaxation exercises, can be helpful, but many of us find that they're not enough. Downtime is also a crucial part of stress relief, as we'll consider later in this chapter, and most of us would benefit from more of it. But as I've found for my patients and for myself, we can experience stress no matter what we're doing. True rest is less about doing or not doing and more about being.

Be: Lay It Down

Years ago, I worked for a difficult boss who continuously demanded that employees prove their worth. Those were exhausting days—it's hard to feel at rest when you're always being told you're not doing enough. What a relief it was to leave that job and the daily stress it brought. But I came to realize that I was perfectly capable of generating my own stress from within.

Most of us bear the constant weight of fearing that we won't measure up, even without an overbearing boss. In a subtle voice we may not notice, we're critiquing nearly everything we do: "Is it good enough?" I've caught my own mind's sotto voce scrutiny even while doing things I love, such as cooking: "Are you sure you're squeezing enough water out of the shredded potatoes?" it whispers as I'm making latkes. Every task can feel like a test of self-worth and an opportunity to fail and disappoint someone—our employer, our kids, our parents, our spouse, God, our higher self. That is a heavy load. No wonder we feel stressed all the time.

Think about any feelings of inadequacy that you might be carrying. How often do you feel that you're not measuring up? What self-critical thoughts do you tend to have? Consider how these mental patterns contribute to your daily stress.

It doesn't matter how much time off we take if we don't lay down our fears about not being enough. I have taken my stress with me on vacation, worrying that things would fall apart while I was away and dreading all I'd have to do when I returned. I didn't trust that I could let go, fearing a phone call that a patient was in crisis. I was sure things would go sideways without my continual attention and emotional involvement. There was no break from striving to make sure everything would work out.

Even my meditation practice brought little rest when I approached it as another item on my to-do list and another test that I could pass or fail. Finally, I realized that mindful trust is the heart of rest. I began to trust that I could stop striving to avoid failure or to be something more and that I was allowed rest in being who I already was. That was the key I'd been missing as my own stress took its toll.

My patient Alex found this deeper form of rest. Like so many working parents, she felt that she was failing both at work and at home and that everyone was disappointed in her. She'd developed nightly insomnia and had the "tired and wired" look I'd seen in many of my patients. Alex spoke wistfully of her fantasy to walk away from the long hours of her high-pressure job and live a simple life and to share more time with her partner and their two young kids. She longed for the rest that her mind and body needed, and she hoped that mindfulness could help.

Alex worked with me for several weeks on practicing mindful presence and acceptance, both through meditation and in her daily activities. She learned to focus on what she was doing when her mind was drawn to the uncertain future and to accept herself and her situation—including her very human limitations.

I was struck by the changes I saw in Alex when I ran into her a couple of years after our last session. She had continued to practice the meditation and everyday mindfulness that we had started together, and she exuded a lightness I hadn't sensed even at the end of our work together. She had even coordinated a training in mindfulness at her workplace, and she was sharing the practices with her wife, as well. Together, they were finding more harmony in their marriage. Alex still experienced the stresses that came with her demanding job, but she was no longer preoccupied with the possibility of failure.

The fundamental elements of mindfulness that we've explored can facilitate rest.

As we focus on what's right in front of us, we let go of the future threats our minds create.

When we embrace reality, we release the exhausting struggle to make things go our way.

From this grounded place, we can decenter from difficult thoughts and emotions and observe them from a distance. We can even decenter from stress itself and recognize that although we're *experiencing* stress, there is an observer within us that is perfectly at peace.

I felt the same sense of lightness as Alex when I finally laid down the mental and emotional weight I'd been carrying. It had been with me for so long that I thought it was a part of me. I realized I didn't have to be so hard on myself and that my worth was not based on my productivity or success. The relief I found was closer than I knew. It didn't depend on a vacation or on a special way of breathing. It's available in every moment—even when we're active.

WORK WITH EASE

My extraordinary voice instructor, Diane Gaary, has often encouraged me to consider whether I can "do less." She doesn't mean taking time off but rather finding more ease in whatever I'm doing, whether speaking, walking, writing, or anything else. With mindful presence, this ease is always available. We can learn to employ only as much effort as we need for whatever we're doing, without additional strain.

Witnessing this ease in action can be mesmerizing—the relaxed wrist of a violin soloist, the fluid form of an Olympic swimmer, the unhurried flow of a clear stream. Effort is completely focused on the work at hand rather than on trying to force a desired outcome. The hawk circling outside my office window years ago captivated me as it embodied the ease I was missing. The best thing about this type of rest is that we can find it no matter what we're doing—at work, on vacation, during pauses in our activities. I'm doing less now as I write this than I was a moment ago, just by realizing I could.

Maybe even in this moment as you're reading, it's possible to do a little less. Release something you don't need: unnecessary physical tension, excess effort, background worry, a baseline assumption of failure. Allow the feeling of lightness that comes when you're no longer trying so hard.

Establishing mindful awareness and ease is essential for finding wholehearted rest. Without a foundation in *being,* our efforts to relax can feel long on effort and short on relaxation, as if stress management were one more thing to achieve. Mindfulness helps us let go of our task orientation and simply tune in to this moment. With that greater awareness in the present, we can recognize stress's signature—frantic thoughts, tight stomach, clenched jaw, the quality of energy coursing through us. And we can choose thoughts and actions that allow us to let go of stress and find rest.

Think: Let the Mind Support Rest

The stories we tell ourselves often prevent us from allowing ourselves to rest. Many of these stories come from implicit core beliefs that are so familiar, they feel like bedrock truth. A common thread running through these thoughts is that we somehow have to make up for a deficit.

IDENTIFY THE STORY

Alex's story was "I have to do it all"; rest felt irresponsible when there was so much to do. My story was "I'm not allowed to stop striving." I constantly had my foot on the gas, driven by fearful thoughts about running out of time—that I would reach the end of the day, or the month, or my life, and realize with regret that I hadn't done *enough*.

If you're struggling to get the rest you need, start listening for the story. Common ones include the following:

Rest is for the weak (lazy, entitled).
I need permission to rest.
More rest means less productivity.
I don't deserve to take a break.
I'm not doing enough.
If I'm not stressed, I'm not working hard enough.
I have to make everyone happy.

Choose a different starting point—an assumption of being enough, while trusting you'll have everything you need—to transform your relationship with rest. This mental shift can also relieve the pressure of feeling that you don't have enough time.

BEFRIEND TIME

Our relationship with time is central to our experiences of stress. My wife and I worked at a summer camp years ago where the staff took turns doing dishes for scores of campers. Some of the staff seemed desperate to get the job done when it was their turn, throwing things around and yelling at people to go faster as they scrubbed pots and pans and loaded the dishwasher. The underlying message that "this is taking too long" created a frenzied feeling of stress in the kitchen.

The relentless feeling that everything is taking too long has often been my Achilles' heel. It was exhausting to feel I was always on the clock no matter what I was doing—laundry, cooking, exercising, writing. Even setting the oven used to be an ordeal, with each press of a button raising the temperature by five degrees. It felt as if I were standing there *forever* if I had to set it to 450 degrees. Finally, one day I asked myself: How long does it *actually* take to get to 450? (Drumroll . . .) Seven seconds. I had to laugh. *That* was what I'd been making myself miserable about? Surely my well-being was worth more than seven seconds.

Many of us feel a constant pressure to go faster, as if time were our enemy. Everything takes too long when we are rushing through what we're doing to get to the next thing. When we insist something is taking too long, we're fighting against time, which is a battle we never win. Wishing things were taking less time won't make them go faster and will sap the enjoyment out of any activity.

Allow one activity today to take as much time as it requires. Let go of the urge to rush through it to get to the next thing. Remind yourself as often as you need to that the clock is not your enemy, and just be in the task for as long as it takes.

We can also beware of the common error of believing we can squeeze "one more thing" into too little time, such as putting in a load of laundry before leaving for an appointment. The added time stress usually isn't worth the small gain in productivity.

When we tell ourselves better stories, we're better positioned to plan behaviors that allow us to rest.

Act: Rest from Doing

Stress is often self-perpetuating. The feeling of threat triggers a reflex to *do more*, even when what we really need is a pause. It's like a faulty thermostat: instead of turning off the heater as the room gets hot, it turns it up. The result is a runaway cycle of busyness and stress.

When I was overwhelmed by work and other commitments, no amount of relaxation practice or healthy thinking could counteract the excessive stress I was taking on. I kept trying to open windows to cool things off when what I needed to do was turn down the heater. Although it's true that we can be present and find ease in anything we do, at some point the sheer volume of too much activity will preclude rest.

We're not machines that can work nonstop without losing our equilibrium. "Our activities create something like a centrifugal force," wrote David Steindl-Rast in *Gratefulness, the Heart of Prayer*. "They tend to pull us from our center into peripheral concerns. And the faster the spin of our daily round of activities, the stronger the pull."[1] It's easier to be present when we're not whirling about.

A crucial part of my healing from excessive stress has been periods of nondoing that allow my mind and body to relax and provide rest for my overtaxed nervous system. Just as stress demanded that I do more, finding rest leads to more rest. In these pauses, I can find the mental and emotional space that brings me back to my center and helps me see through my assumptions about needing to do more.

GET AWAY

One of the most obvious ways to take a break is to go on vacation. Research confirms the many benefits of vacations for our health and well-being: more energy, a brighter mood, and feeling healthier, along with less tension and fatigue.[2] However, these benefits tend to be short-lived, as you probably know from personal experience; virtually all of the physical and emotional payoff from a vacation disappears within a week or two of returning home.[3]

These research findings don't mean that the benefits of vacation are negligible. One of the best aspects of getting away is having the space to find a new perspective on our lives, as with being able to see a mountain more clearly from a distance. We might realize that the way we're living isn't working for us and choose to make changes when we return that last far longer than the direct effects of taking time off. For example, it was only on vacation that I realized the changes I wanted to make with my screen time, as I described in chapter 6.

Nevertheless, we can't bank our peace of mind on a few weeks of annual vacation. Even four weeks per year—more than most of us take—leaves forty-eight weeks to feel harried and stressed out. I know that for myself, a week of vacation each summer wasn't enough to stem the rising tide of burnout and sickness. Vacation itself can be exhausting, and we often have a pile of work to catch up on after being away. Clearly, we need ways to address our stress that don't depend on infrequent vacations.

REST DAILY

For a long time, I was amazed at how little downtime I had each day. It wasn't as if I was deliberately planning to be busy. But even when

I wanted to work less, my schedule always seemed to fill up. I finally realized that my hope for rest wasn't enough. I needed to dedicate specific times for nondoing or my days would fill up with activity like crabgrass spreading through a garden bed.

I finally accepted that I had to *plan* to be less busy—just as I plan to keep the weeds in my garden at bay—and make a commitment to protecting my schedule. My own Think Act Be plan included setting aside time for daily self-care: to read a book for pleasure, sit in my garden, talk with my kids, do nothing. These periods of calm and relaxation invited a quality of presence that was inherently stress-reducing.

Begin to plan your day as if you care about the person you're planning for. Carefully examine your schedule for tomorrow and notice whether there's any optional stress built into it. For example, have you agreed to do something that you could opt out of? Also see if you've allowed for any downtime between activities and time for essential things such as sleep and exercise. You don't have to revolutionize your calendar overnight; aim to make one improvement for tomorrow that makes you more excited about the day ahead.[4]

From a quieter, more centered place, we can reconnect with the spiritual part of ourselves that can easily get pushed aside when stress is high. The most effective approach is to make restful activities part of our routine, such as going for a walk every morning or having lunch with a friend on Fridays. That way, we don't have to decide every time whether or not to do them. Part of my own recovery has included short sessions of yoga that bracket my day, once before breakfast and again before bedtime.

Restful breaks can be brief; taking just fifteen minutes at lunchtime to go for a walk or do relaxation exercises significantly lowers stress and fatigue—not just in the moment but at the end of the workday.[5] I've often found a mental reset at lunchtime by taking five minutes to walk outside and check my mailbox. By making a small investment in our well-being at midday, we can finish the day feeling more energetic and relaxed. Importantly, the benefits of the lunchtime breaks come not just from pausing in our work but from doing things that fill us up.

REST IN ALIGNMENT

Many of the people I've worked with have struggled to find restorative downtime. David, for example, would come home exhausted from work and school and watch YouTube videos for most of the evening. He was relaxing in a way, but he felt just as empty at bedtime as when he got home. Eventually, David realized he was starved for human connection.

Try this approach when you're feeling overly stressed by all you have to do. *Be*: Pause and come back to yourself, using a single slow breath to find your center. *Think*: Then ask yourself what the story is, such as "I can't get it all done" or "I can't afford to take a break." Are your thoughts completely true, or is there a more realistic alternative? *Act*: Finally, see what form of rest you can offer yourself, like a five-minute walk around the block or a relaxing shower. If you can't rest immediately, can you plan to do it soon as something to look forward to?

True rest is found in giving our minds, bodies, and spirits what they need. Sometimes we need to rest by lying around and reading all weekend or by taking a nap. At other times we need to get outside and move our exercise-starved bodies. Sometimes rest means being alone; David found that he needed to rest from solitude by getting together with friends after work, which he found much more fulfilling than his old routine. Mindful awareness helped him to listen for what he really needed in the evening and to respond accordingly.

Tending to our thoughts and our actions can go far in reducing our stress and help to foster ever deeper experiences of being at rest—even from wanting more.

REST FROM WANTING

We're strongly conditioned to think of ways to improve our situation by finding a bit of pleasure or removing discomfort: instinctively grabbing something to eat, flipping through television channels, scrolling through social media—anything to remove the vague sense of dissatisfaction we feel so much of the time. But the constant desire for things to be better puts us at odds with reality and precludes peace. Even the drive to get rid of stress can be a state of nonrest, creating a stress of its own.

We often aren't aware that this dynamic causes stress because it's so habitual. An obvious example from my own life was found in my relationship with alcohol, which I relied on for years to enhance my life: to relax, make socializing easier, add to the experience of watching sports. But the acquisitive drive behind my drinking always brought a measure of stress as I chased an elusive feeling of *just right*.

The rest we long for does not depend on enhancing what's happening. It's found in resting from the belief that we need anything more than we already have. The Trappist monk Thomas Merton described this rest as "beyond all desire, a fulfillment whose limits extend to infinity."[6] We can stop reaching for the things we're attached to, coming to rest in this very moment and releasing any expectation that things should be different or better. Rest is found in that acceptance.

Mindful acceptance is what helped my patient Alex to lay her burden down. Eventually, she came to see that she'd been at war with herself. Through mindful awareness, she learned to make friends with who she was and with the messy uncertainty of her daily existence. *That* was the home she longed for—a home where she felt truly welcomed and at peace and reconnected with her spiritual core. As she came home to herself, she found rest in the life she was already living.

Our spirits dwell, only and always, in the utter simplicity of the present moment. Finding spiritual connection helps us to balance doing and nondoing and to question the thoughts that fuel our stress. What could compel us to constantly strive for more when we no longer need to run from our fears or prove our worth? We find rest in knowing we already have everything we need.

Rest is the cornerstone of our well-being and self-care. It allows us to connect with ourselves just as we are and to feel at home in each moment of our lives. And, as we'll see in the following chapter, it's an essential part of fostering a loving relationship with our bodies.

LOVE YOUR BODY

Your relationship with your body is a fundamental part of connecting with yourself. We all know that taking care of our body is a good idea, and yet we often struggle to get adequate sleep, exercise consistently, and eat a healthy diet. This chapter shows you how the Think Act Be approach provides leverage to follow through on your intentions in these areas. As you offer your body loving care and connection, you can feel more at ease in your physical home.

* * *

My health started to fall apart as I approached my fortieth birthday. First, it was recurrent laryngitis and vocal strain, which I treated with ten to fifteen cough drops per day. I assumed my body was being unnecessarily disagreeable and never considered that it might be trying to tell me something important.

The message soon got louder and more insistent. "I could use a little help here," my body announced when I didn't have the stamina that had always carried me through my morning runs. I chalked it up to turning forty and the inevitable decline of age. And then suddenly it was screaming, "We have a problem!" My sleep fell apart. My memory was spotty. I felt confused. I couldn't complete more than a minute or two of a workout before crushing fatigue set in. I used to take the subway stairs two at a time, and now I was taking them slowly one by one, an aching heaviness in my thighs, and stopping at the top to catch my breath. All the while, it was becoming more and more difficult to speak.

At the time, I didn't understand what was happening. My body had done everything I'd asked it to for the first few decades of my life—from running and cycling to long workdays and too little sleep. The only time I paid much attention to my body was when there was a temporary problem, like getting a cold or throwing out my back. As my illness intensified over the next four years, I cursed my body, feeling it had betrayed me.

Eventually, I realized that I was the one who had betrayed my body, treating it as a distant afterthought and taking for granted that it would serve my needs. I'd demanded that it take care of me even though I was neglecting it, as if I were cursing a struggling garden that I never tended.

Through my illness I learned that everything we value, from our jobs to our sensory experiences to our deepest relationships, depends on caring for these bodies that are entrusted to us. How can we treat our body as something worth taking care of?

Mindful presence is essential for taking care of our body. When our mind is in the past or the future, we lose connection with our physical self, which exists only here and now. Rushing to get to the next thing actually puts us at odds with our body; if we believe we need to be *there*, then our body *here* is a problem. In contrast, being present grounds us in physical reality and allows us to foster an intimate, caring relationship with our body. Taking care of our body feeds back into our mindful awareness, strengthening our connection with the present (see figure 5).

Figure 5

Let's explore three major ways to care for and connect with the body: sleeping, moving, and eating.

Sleep Soundly

Sleep is the first pillar of self-care. For a long time, I shortchanged my body's need for sleep, begrudging those seemingly empty hours. But even though sleep is a state of nondoing, it's anything but wasted time.

I learned the real value of sleep through several years of chronic insomnia. Falling asleep was easy, but staying asleep was not. I was awake at two or three o'clock most mornings, often getting up to start my day rather than tossing and turning in bed. It was a lonely existence to be awake and active while the world around me slept. I moved through countless days in a disconnected haze, looking forward to being back in bed but fearing another sleepless night.

Few things are more important for our well-being than good sleep. Countless processes depend on it, including memory, concentration, energy level, immune function, tissue repair, and hormone regulation. Good sleep helps prevent anxiety, depression, and alcohol problems[1] and can make you a more productive worker, a safer driver,[2] and a better partner.[3] Chronic sleep problems are frustrating and dispiriting and can make us feel like just a shell of our full selves.

How can we experience sleep that not only is sound but also nourishes our spirit? The Think Act Be principles that underlie cognitive behavioral therapy (CBT) for insomnia provide exactly what we need to enter into the hallowed space of sleep. The first thing to do is to *not* do.

BE: LET GO

Sleep is a nightly invitation to mindful presence, sitting as it does between the day you've just had and the day you're about to have—between memory and fantasy. For a few hours, there is nothing else to do and no need to be anywhere other than where you are. Sleep requires you to do exactly one thing, though "nondoing" is more precise, since it's one of the very few activities in which more effort yields worse results.

Rather than *trying* to fall asleep—which suggests that sleep is under our deliberate control—we release into it. Our only task is to surrender. Falling asleep is a kind of letting go and an act of faith. As the Franciscan priest Richard Rohr wrote, quoting Angeles Arrien, "Every night we practice letting go when we release ourselves to sleep and the mysterious place of dreams, trusting that we will return."[4] Trust is essential for sleep, since a feeling of threat is sure to keep us awake. We also trust that we can pause from thinking and doing and rest in a state of being.

Research shows that few things are more valuable than mindful acceptance for avoiding frustration when our sleep is interrupted by insomnia.[5] Part of acceptance is embracing uncertainty by trusting that sleep will come eventually when we offer our mind and body the right conditions. Not surprisingly, I've been sleeping better since I started bringing mindfulness to bed with me. When I accept whatever happens each night, I can let go of the demands and expectations that chase away good sleep.

It's easier to accept uncertainty about our sleep when we bring a "beginner's mind" to it, approaching sleep as if for the first time (as I described in the Raisin Exercise in chapter 5). Instead of assuming that we know exactly how the night will go, we can open to each night for what it is. Sometimes we'll sleep poorly; on other nights, our sleep will be easy and sound.

THINK: CONSECRATE YOUR MIND

Early in her treatment for insomnia, Janine lamented to me that she wished she had an "Off" switch for her mind. It was hard enough during the day to let go of distressing thoughts about real and potential problems; at night, it was even harder to stop grinding the wheels of worry, with no distractions to interrupt her fears.

Often, the difficult thoughts that ran through Janine's mind were about sleep itself, telling her stories about how terrible the night was going to be and how she would suffer the next day. "The most sleep I can get now is four hours!" she would think. "I'm going to be a wreck tomorrow."

If she tried to argue with those scary thoughts or make them stop, they would just get louder. During her treatment, she learned instead to practice decentering from them, seeing the thoughts as mental chatter rather than facts. Over time, she could observe those stories from a little distance without getting lost in them. "Maybe they're true, but maybe they're not," she could tell herself. For example, she realized that even though a bad night's sleep was very frustrating, the next day was never the disaster she feared it would be.

Practice questioning the stories your mind tells you at night. Begin by noticing that you're having thoughts that might not be true. Then see whether there are any alternatives that might be more likely or accurate. Let go of any attempt to force yourself to believe these alternatives; instead, start to give less weight to the mind's mental chatter.

When my insomnia was starting to improve, I would often have intrusive thoughts when I woke up in the early morning hours: "What if I'm up for the rest of the night?" Rather than wrestling with that possibility, it was helpful to open to it: "I might be, and I'll have to deal with it if it happens." Accepting the unknown makes it easier to get more sleep, since we aren't trying to force ourselves into unconsciousness.

ACT: CONSECRATE TIME AND SPACE

Behavioral principles help us to align our actions with our body's two main sleep drives: our internal clock and our hunger for sleep.

Sleep-related processes such as hormone release and temperature regulation follow a roughly twenty-four-hour cycle known as our circadian rhythm. For example, levels of melatonin, which signal our bodies that it's time for sleep, rise shortly after the sun sets and decrease as morning approaches. We sleep best during the phase of our cycle that our bodies associate with sleep.

The other crucial factor—sleep hunger—is like our appetite for food; the longer we go without sleep, the more we crave it. Sound sleep is more likely when we've worked up a good appetite for it, which helps us enjoy a full and satisfying "meal" of sleep. You'll sleep best when your sleep hunger and your circadian rhythm are aligned—that is, when you've been up long enough that you're ready to sleep and your internal clock says it's bedtime.

SCHEDULE TIME IN BED

The foremost principle for aligning your two sleep drives is to follow a schedule that matches your body's sleep needs. A consistent sleep schedule is crucial for building a strong circadian rhythm so that your time in bed is synchronized with the physiological processes of sleep. You also need to spend the right amount of time in bed.

Before I developed insomnia, I routinely deprived myself of sleep as many of us do, relying on caffeine to overcome fatigue during the day. Now I'm typically in bed for about seven hours—from ten o'clock in the evening to five o'clock in the morning—because I tend to sleep about six and one-half to seven hours per night.

Importantly, too much time in bed can be just as detrimental as too little. If we're in bed for longer than we're able to sleep, we'll have more frustrating awake time, which can turn the bed into a place of anxiety instead of rest. Aim for the sweet spot where you're adequately rested and are asleep for most of the time you're in bed.

PREPARE THE BODY AND BEDROOM

Good sleep actually begins with our daytime activities. We can prepare the body for sleep by exercising during the day; studies show that both aerobic exercise[6] and weight lifting[7] can improve nighttime sleep. Some people find that exercising in the evening interferes with sleep, but some research has actually found that it's more beneficial to work out closer to bedtime.[8] As always, it's best to pay attention to what our own bodies are telling us.

We'll sleep better when we avoid using alcohol to fall asleep, which hurts our sleep quality and makes us wake up more often. A big meal shortly before bedtime can also disrupt the quality of our rest, as can caffeine. For some people, even a morning cup of coffee or tea can affect their nighttime sleep.

We can also make our bedroom conducive to sleep by keeping it dark, quiet, and cool; having a comfortable mattress and pillow; and choosing nice sheets and blankets.[9] It's best to reserve the bedroom for sleep-related activities by leaving work and digital devices elsewhere; that way, our brains will equate the bedroom with sleep.

WIND DOWN

I used to check my work email right before bed, which often filled my mind with anxious thoughts about potential problems. Political news, intense movies, or exciting books can similarly wind up the mind when we want to wind down. We sleep best when our evening activities align with the body's preparation for sleep, which begins well before we lay our head on the pillow. Lower-energy activities help to shift our brain waves toward the less noisy patterns that are consistent with falling asleep.

An evening winding-down routine was a key part of relieving my insomnia. In the hour before bed now, I typically do deliberately calming activities such as taking a shower or bath, reading something I enjoy, and doing a few minutes of bedtime yoga. These activities prime my brain and body for sleep and provide a smooth transition from doing to nondoing.

Reflect on how you tend to spend the half hour to hour before your bedtime. Do your pre-bedtime activities tend to wind you up or help you to move toward sleep?

Although a winding-down routine can be very beneficial, it's often hard to release all the stress that has built up throughout the day in the hour before bedtime. My stress accumulation typically started in the morning with the physical toll of my illness and constant worries about how the day was going to go. Often, I was still in a hyperaroused state when I went to bed, no matter how much yoga or breathing exercises I did in the evening. As Saundra Dalton-Smith wrote in *Sacred Rest*, "Good-quality sleep trickles down from a life well rested."[10] When we learn to recognize unreleased stress in the mind and body (see chapter 8), we can practice letting go of tension throughout the day, making it easier to sleep well at night.

Like most of the people who came to me for insomnia treatment, Janine found relief through the simple principles of Think Act Be.

She was soon able to fall asleep more quickly and sleep more soundly. She ended up getting more sleep while actually spending *less* time in bed; less awake time turned her bed into a place of rest instead of frustration. When Janine's sleep improved, everything changed. She felt not just less exhausted but more optimistic. Her anxiety also diminished, and she found her sense of humor again.

For many years, I had a one-dimensional view of sleep as either good or bad, and my sole focus when addressing sleep was on trying to make it better. But it took me longer to recognize that sleep isn't just a means of mental and physical rest—it's a daily ritual of cleansing and renewal and the foundation for being present in our waking lives. By preparing our minds, bodies, and spirits, we can discover the best and most mysterious aspects of our nightly dip into unconsciousness. With mindful awareness, sleep becomes a deep spiritual practice.[11]

DIVINE SLEEP

In *The Varieties of Religious Experience*, William James cited the sleep experience of a Madame Guyon:[12] "My sleep is sometimes broken—a sort of half sleep; but my soul seems to be awake enough to know God, when it is hardly capable of knowing anything else." For Madame Guyon, sleep seemed to open a portal to the divine.

Sleep has a prominent place in many spiritual traditions. Hebrew scriptures feature many instances of divinely inspired dreams, such as when Joseph forecast seven years of plenty and seven years of famine based on the pharaoh's dreams.[13] Early Christian writings also reference divine guidance received during sleep, as when the Magi were warned not to return to King Herod after presenting Jesus with their gifts.[14] Sleep is described as a gift from God in the Psalms: "In vain you rise early and stay up late, toiling for food to eat—for he grants sleep to those he loves."[15]

Hinduism also addresses sleep: "The happiness of deep sleep is greater than all other forms of happiness or pleasure born of sense-contact," wrote Swami Krishnananda.[16] Similar views of sleep appear in writings from Sufism,[17] a form of Islamic mysticism. The Buddha drew a connection between sleep and enlightenment, saying that "the enlightened one . . . always sleeps happily";[18] other Buddhist writings teach that some dreams may be prophetic.[19]

There are descriptions in Jewish mystical traditions of the soul leaving the body during sleep and receiving "revelations that cannot happen . . . when a person is awake."[20] Many of us have experienced revelatory sleep, such as when we wake up with the answer to a problem that we couldn't solve during the day. I felt that my death dream of finding perfect love among the stars revealed something true about the nature of reality. It's mind-bending to realize that our sleeping selves may know things that our waking selves do not.

Sleep offers a nightly opportunity to enter into the realm of spiritual consciousness. By nurturing our sleep, we're feeding our connection with the deepest parts of ourselves. That connection not only can help us experience all the gifts that sleep offers but also can enrich our waking lives.

Practice a ritual that marks the beginning of your sleep period, just as martial artists bow each time they cross the threshold of their practice space. For example, take three slow breaths as you settle into bed, and release all that does not pertain to sleep. Upon reentering the waking world in the morning, take one minute to ground your awareness in your body and start your day from a place of centered presence.

Sound sleep sets the stage for the action of our day—including regular exercise, our next way of caring for the body.

Move as You're Made To

Charles came to my therapy office worn down and depressed after years of stress and long work hours. Part of the behavioral activation we planned together included getting back to the running he loved but had given up because of his work schedule. A few months after our last session together, Charles sent me a letter saying he was still doing well. Although he was careful to let me know that I'd been helpful, he said the greatest healing he experienced came from his runs. Several patients have told me that their best therapy was physical activity such as running, swimming, or cycling.

Physical exercise is a well-established antidepressant,[21] but it gave Charles more than symptom relief. Running felt true to who he was as a physical being with a body that longs to move. He felt like himself again as he connected not only with his body but also with his whole being—head, feet, and heart. Charles had rediscovered a childlike joy in moving.

Kids love to move, but at some point, moving stops being our joyous default. Physical activity becomes compulsory and regimented. I suspect I left some of my love for running on the cinder track as I ran the requisite mile for seventh grade physical education class. My body was just a tool for reaching a target as I focused on the clock for four joyless laps.

And yet we're made to move. Our ancestors couldn't avoid physical movement, since their survival depended on hunting and gathering their food. Now it's common to spend most of our waking hours sitting.

We might not even realize how sedentary we are. Before my sickness, I thought I was living an active lifestyle by doing a short workout in the morning and cycling for five minutes to and from work—ignoring the ten hours I spent sitting in my office and the hours of sitting at home. Sitting became such a strong habit for me that I wouldn't even want to get up to refill my water glass or find a book on my shelf.

As my body grew weak and stiff from disuse, it didn't feel good to move, and I was more injury-prone. My hips were chronically tight, so walking was uncomfortable, and my back would often seize up when I stood after sitting for any length of time. I felt so much older than my forty years. These aches and pains further sapped my motivation to move, leading to a viciously sedentary lifestyle.

I'd given up my morning exercise in the early stages of my illness as exhaustion set in, and I gradually quit cycling and lifting weights, too. The lack of movement fed my growing depression and frustration with my body. To be honest, I was envious of Charles's runs. The physical benefits of exercise are no secret, and it can also lower anxiety[22] and relieve depression.[23] Physical activity supports other parts of a healthy lifestyle, too, such as better eating[24] and sounder sleep.[25] And yet we often struggle to move our bodies.

How can we rediscover the joy in moving our bodies every day? The principles of Think Act Be can help. We start with *being*, connecting with our bodies in the present; that connection makes us more inclined to do the work of caring for these temples of the mind and spirit.

BE: CONNECT WITH THE BODY IN MOTION

In the past, my workouts were a lot like whipping a horse around a track; my body was there to serve me, with no sense of collaboration.

The exercise that was to be part of my self-guided treatment would have to be based on a close partnership with my body.

Tuning in to my body helped me to realize what it needed. It wasn't time for high-intensity interval training or spin classes. I made walking and yoga my primary modes of exercise, along with short sessions of strength training. After a particularly poor night of sleep, I would scale back the intensity of my workout to match my energy level instead of demanding that my body perform as if it were well rested.

During this period, I discovered joy in mindful movement, particularly through yoga. I'd never known how good it could feel to sense my body as it moves and breathes. Any conscious connection with the body in motion can be astonishing, from the involuntary movements of the breath to the dexterity of the hands as they open and close. (Try it now if you like.) There's also joy in watching others move. I've been mesmerized many times in therapy sessions by the gestures of a patient's hands as they talked. Human movement is inherently poetic.

From a foundation of mindful presence and awareness, we can attend to the thoughts that can interfere with exercise.

THINK: REMOVE MENTAL BLOCKS

The obstacle to movement is often in our minds. When I've fallen into periods of inactivity, I've caught my mind making many classic thinking errors. The biggest one is the "I'll do it later" fallacy. I knew very well that if I didn't exercise in the morning, there was a 99 percent chance I wouldn't get to it that day. But if I was in a hurry to get started on work, I could trick myself into believing I would have the time and inclination later.

It's also not very effective to try to "should" ourselves into moving. Shoulds can actually lower our motivation because they evoke resistance. When anyone—including ourselves—tells us we should do something, a part of us will rebel against what feels like an onerous obligation. It's more helpful to focus on why we *want* to get moving,[26] such as the joy we might feel as we do it.

I've often tricked myself with emotional reasoning, too. If I don't feel like moving, I'm liable to think I shouldn't do it. But our feelings often follow our actions, so once we start physical activity, motivation tends to follow. Emotional reasoning is often compounded by fortune-telling, when I tell myself I'm going to be miserable while I'm working out. This prediction is usually based on how I feel at the beginning of the workout, as when I think I'll feel cold the whole time I'm in the pool. In reality, the initial jolt of cold quickly gives way to the pleasure of gliding through the water.

Also beware of black-or-white thinking, which says that if you can't do your whole workout, you might as well skip the whole thing. Something almost always beats nothing when it comes to exercise.

When it's hard to get moving, see whether the mind's stories might be getting in the way. Are your thoughts sapping your motivation to exercise? Start to replace unhelpful thoughts with more realistic ones; for example, counter black-or-white thinking by reminding yourself that any movement is beneficial.

As you challenge unhelpful thoughts, you'll be clearing the way for action.

ACT: MAKE IT FUN AND EASY

I still remember the best workout I had in the summer of 1999—two hours of folk dancing when I was working at a summer camp in

Maine. It was so much fun I didn't even notice how strenuous it was. Unfortunately, we often have the opposite experience with exercise, feeling all the pain and none of the joy. It's really hard to commit to lasting changes that feel like punishment.

Enjoyable movement, on the other hand, can feel more like dancing than exercise. As we're building a more active lifestyle, we can ask ourselves what type of movement would bring a smile to our face and help us feel alive. Exercise is just as good for you when it's not punishing, and you're much more likely to stick with it.

Look for activities that bring a relatively immediate payoff rather than promises like "a longer life" or "a stronger heart." Rewards are much less inspiring when they're beyond our mental horizon. More immediate rewards might include exercising with other people or doing movement that's a good fit for your body. I always dreaded running but rarely had to convince myself to get on my bike.

Gradual changes in our movement routines are best. We're naturally drawn to thirty-day challenges and dramatic reversals in our exercise routine, but those approaches don't often stick. Once we miss a day, it's easy to fall prey to the "what the hell" effect and drop the whole thing. Instead of planning to exercise seven days a week, for example, we can aim for twice a week. This approach is also better for the body, giving it time to adapt to new demands so we don't hurt ourselves.

Part of building gradually can include breaking down our exercise plans and committing to one small step at a time. If we're going for a run in the morning, we might get out our shoes and shorts the night before. That way, we won't use up any of our early morning motivation trying to find our gear. In addition to lowering the cost of moving our bodies, we can raise the cost of bailing on our workout by increasing our accountability. I know I would have skipped many more runs if I hadn't bound myself to the mast by having a running partner who was expecting me.

FALL IN LOVE AGAIN

Your body wants to move in all the ways it can, whether stretching or walking, squatting or lifting. It's easier to get moving when you fall in love with physical activity and look for ways to move rather than for ways to avoid moving.

Find small, nonexercise ways to use your body throughout the day, such as getting up more often to refill your water container. Reclaim movement as the default—not because it's good for your body, prevents dementia, boosts your mood, and improves your sleep. Move for the sheer joy of moving, letting your body do what it's made for.

If you find yourself struggling to exercise, come back to connection with your body. Tend first to that primary relationship. As we ground our awareness in the body, movement becomes meditation in action. We discover gratitude, even reverence, for the inspired matter that makes up our physical self. Every movement becomes an act of praise.

Sit comfortably, letting your hands rest in your lap. As you breathe in, reach your arms out to the sides, palms down. Pay close attention to the sensations of moving. Breathe out and bring your palms together in front of your heart. Inhale and ex-

tend your arms again, and then exhale and rest your hands in your lap. Repeat twice more as you continue to tune in to your body and breath.[27]

Movement and diet are intimately connected as the food we eat becomes energy and action. Every day provides multiple opportunities to love our body through what we put in our mouth.

Eat for Life

When I was in full-time clinical practice, I rarely took a proper lunch break. I resented the time it took to eat, as if it were intruding on what I *should* be doing. On most days, I would eat lunch at my desk as I wrote my notes or caught up on email. If I was particularly rushed between sessions, I had to choose between a snack and a quick bathroom break. *Or did I . . . ?* My solution was often to shove a handful of nuts into my mouth and chew them up during my trip to the bathroom. Besides showing a lack of self-respect, that's just gross. Eating felt purely utilitarian, with no connection to the food that was animating my experience.

Much of the time, we treat eating like a throwaway activity, scarcely realizing what we're putting in our mouths. Many of us also struggle with eating too much and too many unhealthy foods. We might be dissatisfied with our nutrition habits but struggle to make lasting changes.

How can we eat in a way that nourishes our mind, body, and spirit? Eating can be an exercise in joy and a celebration of being alive when we're present for our meals.

BE: FEED CONNECTION

Lettuce doesn't usually make me emotional, so I was blindsided when a single leaf brought tears to my eyes at a dinner party years ago. I'd had thousands of salads in my life, but none like the baby greens the host had just picked from the garden. She offered me a sample before dinner, and it felt as if I was tasting life itself.

Our food doesn't have to be freshly picked from the garden for us to connect with it; any meal can be an invitation to celebrate our existence through the very stuff that keeps us alive. It doesn't take much to reestablish connection with our bodies and our experience. We can make it happen just by taking a slow breath or two before we eat and acknowledging that we're about to feed our body.

When you sit down to a meal, take three slow, calming breaths. With the first breath, bring your mind back to your body, taking stock of any sensations you notice. With the second breath, open your awareness to your surroundings, including any people you're eating with. On the third, look at the food you're about to eat, noticing the colors, textures, and aromas.

It's easier to be present when we sit at a table meant for eating, put away other activities, and dine with others if possible. Sharing food connects us with other people; the word "companion," after all, comes from the Latin roots for "with" and "bread." Breaking bread together underscores our shared humanity. Being present and relaxed also allows for good digestion, which is why the calming parasympathetic nervous system is called the "rest-and-digest" system.

With mindful presence, we become more aware of how our food choices affect the way we feel. Years ago, I enjoyed the best Italian hoagie I'd ever had. About an hour later as I tried to work, I felt as if I'd been drugged. I couldn't even keep my eyes open. I'd had similar experiences many times after eating giant bagels or mounds of pasta, but this was the first time I connected it to all the refined carbohydrates I'd just eaten. It wouldn't have been hard in principle to see the link between certain carbs and my postprandial lethargy, but I was well into my thirties before I recognized it.

The mental health field similarly has been slow to recognize nutrition as a key factor in our well-being. I never asked about my patients' diets until the past few years because I assumed it had little relevance to the issues we would be working on. But a growing number of studies have shown that our mental and emotional wellness are closely linked to the foods we eat.[28]

There's no shortage of diet options to choose from and no clear consensus about what the human body requires. However, most diets agree on one basic premise: the most nourishing options resemble their original form, such as vegetables, fruits, nuts, beans, and fish. Practically all diets also agree that highly refined foods such as sugar and white flour don't offer our bodies what they need.

My health crisis compelled me to make improvements in my diet, which have been one important part of my healing process. I realized that my relationship with food was fraught with anxiety and that I routinely overate as I tried to satisfy a persistent craving for more. What I settled on for myself is what I recommend to my patients when we discuss diet: *listen to your body*, and bring greater awareness to how food and drink affect you.

THINK: MIND YOUR FOOD

Our minds often lead us to eat things we're trying to avoid, but they can also be powerful allies in improving our diet. Jim found in our work together that permission-giving thoughts were driving his late-night eating. "You've been good all day," his mind would tell him. "You deserve to treat yourself!" When Jim identified those thoughts, he realized they were just stories driven by his cravings for foods that weren't good for him. He found it helpful to tell himself, "I deserve to feel good after meals."

We can also catch our thoughts telling us to eat more than we need. I believed, as many people do, that eating enough at a meal meant being *completely full.* But that mindset often led to my feeling uncomfortably stuffed shortly after a meal. A more useful question is "Am I still hungry?"[29] That simple reframing has helped me listen to my body's needs and avoid eating to the point of discomfort.

ACT: WORK WITH YOURSELF

Behavioral principles can also help us eat well. A few years ago, I would routinely have a bowl or two of cereal shortly before bedtime, which led to indigestion and interrupted sleep. When I decided to break the late-night eating habit, I found a simple trick: brushing and flossing my teeth immediately after dinner. I wouldn't want to have to repeat my tooth-care routine later on if I ate again, which took having a midnight snack off the table.

This little habit relies on a big *act* principle: *make it harder to do the thing you don't want to do.* Conversely, we can also reduce the time and energy required to access foods we want to eat more of, such as by preparing healthful snacks in advance. We can also look for more delicious ways to prepare the foods we want to eat more of. For example, I barely tolerated brussels sprouts during my first thirty-five years. Finally, I had them roasted in the oven with garlic,[30] and my life has never been the same since.

As with exercise, we're inclined to make sweeping, all-or-nothing changes to our diets. However, it's easy to flip from *all* to *nothing* when we break our self-imposed rules, as when we have a forbidden bowl of ice cream and then figure we might as well finish the whole pint. Radical changes to our diet also make us more likely to crave the foods we're missing, eclipsing the enjoyment we could be finding in the foods we're eating. As we explored earlier with the example of my not-running friend, it's nearly impossible to enjoy an activity that's defined by what we're not doing.

Aim instead for gradual changes if you're working to improve your diet. You can start by making one meal per week a little healthier—maybe adding a bit more of your favorite vegetables to lunch on Wednesdays, for example. Gradual upgrades will also help you to experience the rewards from what you're adding to your plate without being hit over the head by what you've taken away.

Practice caring for your body by making your next lunch a bit better than usual. Imagine you're preparing it for someone you love. Aim to include nice touches like an actual cloth napkin and metal silverware. Schedule enough time to eat your lunch without rushing, and savor the experience of a meal made by someone who cares about you.[31]

We tend to approach food differently when we start from a mindful relationship with our bodies, as if we are feeding a child we love.

We recognize that we're not just filling our stomach but also fueling the thoughts and actions that make us who we are. This heart connection guides our head and hands toward *think* and *act* practices that are essential for aligning our actions with our intentions. Our resulting food choices will truly satisfy our mind, body, and spirit.

A Full Embrace

Connecting with our bodies is like coming home to an old friend. A single breath with awareness of the body invites us back into the ever-flowing stream of mindful awareness. This simple connection can lead to something deeper than we imagine. Joining mind and body in the present opens the door to spiritual connection.

As we spend time developing our relationship with our bodies, it gets easier to love them—not just to take care of them but to embrace them exactly as they are. Loving your body in this way might be a foreign concept; maybe you've never liked the way your body looks, or you're struggling to accept the changes that have come with aging. I certainly don't have the body—or the hair—that I had when I was seventeen.

Maybe you're frustrated by your body's limitations, as I was during my illness. But our bodies need our care and acceptance more than ever when they're struggling. We can't force ourselves to feel loving toward our bodies; as in any relationship, love emerges through spending time together and paying attention. Receiving and working with the bodies we have is a daily discipline in mindfulness.

Spend a few moments feeling the aliveness in your body—the strength and vitality coursing through you, even if your health

is compromised. Notice what your body does for you every day, starting with breath after breath. Consider also the challenges it has come through and might still be enduring.

I didn't realize what a dear friend my body had been for my whole life, giving me everything it had. Even the alarm bells it was sounding were for my benefit, like a smoke detector going off in the middle of the night. My body and I were in this together, and I started to be thankful that it had woken me up.

I continue to face many unknowns with my health. At the time of this writing, I still deal with health challenges on most days, but my health has greatly improved as I've committed to loving the body I have. I've slowly regained strength and stamina, which allows me to keep up with my wife and kids on family walks. My sleep is much better. I've rediscovered my sense of humor, such as it is. Through my recovery, I've found an appreciation for my body that I never had before, even when I had hair and abs and endless energy.

Our bodies beckon us into a relationship and repay our love when we take good care of them. Tending to our bodies' needs also helps us to bring our best to everything we care about, including our relationships—the topic of the next chapter.

LOVE OTHERS

Having strong, healthy relationships is a crucial part of feeling at home in your world. In this chapter, you'll learn how mindful awareness facilitates connection not only with yourself but also with other people; how to change common ways of thinking that can disrupt connection; and everyday actions that can strengthen your relationships. You'll also discover how to be at peace within yourself even when your relationships are strained.

* * *

Tina was surprised by how upset she had been when she saw her dad holding his newborn niece. I'd been working with Tina for several weeks, primarily on the anxiety she'd battled for most of her eighteen years. "What was hard about seeing your dad with the baby?" I asked.

"It's just that I know he never held me like that," she replied. The grief in her words hit me hard. Tina's dad had been absent for most of her childhood, and she had no memories of feeling his affection.

The painful experience of seeing her dad expressing the love she'd craved had opened a door within herself that she hadn't known was there. On the other side was a cavernous sense of loss, and she understandably drew back from it. It felt as if the depth of sadness could swallow her up.

Even as a relatively new therapist, I knew my responsibility was to accompany Tina into those dark and painful places—to walk with her into the rooms that had been barred shut. As I validated her deep sense of loss, Tina was willing to step inside that scary place and feel the unacknowledged sadness she'd been carrying for so long. In our subsequent discussions, she discovered that much of her anxiety came from a core belief that she was unlovable. That realization helped Tina begin to question the fears of abandonment that fueled her anxiety and hurt her relationships.

The quality of our relationships is the single greatest determinant of our mental and emotional well-being—for better and for worse. In one way or another, relationships have been a central issue for everyone I've worked with in therapy. Some came specifically to work on their marriage or their work relationships. Others were still healing from an abusive childhood or a painful divorce. Not uncommonly, a person had finally come to therapy when their anxiety or depression was getting in the way of their relationships, such as contributing to conflict with their partner.

The Best and the Worst

As a therapist, I've witnessed the worst of what we can do to one another. I treated a man whose mom broke both of his arms in fits of anger when he was two years old. I've worked with men and women who felt lonelier with their partner than when they were actually alone. I've seen patients who were battling health issues as I was but who found more anger than understanding from their closest loved ones. I've treated people like Tina who were wounded by a parent's absence. Each of us is shaped by our closest connections.

At the same time, I've seen the best we can offer each other. Many of those I've treated who had suicidal thoughts said a close relationship helped keep them alive. I was drawn back to life again and again through connection to my wife. When I couldn't see any good that I brought to our family, Marcia reminded me that I was still working and parenting as best I could and that our kids felt love from me. When I felt like a lost cause, she reassured me that I wasn't a failure. I didn't believe a lot of what she was telling me, but I knew *she* believed it, and that was enough.

I can't imagine what it would have been like to go to bed alone every night with my fear, grief, and confusion. It's easy to imagine sinking ever deeper into depression and being lost in its depths. It took no imagination during that period to understand how my dad's dad had taken his own life when he was not much older than I was. Without the buoying force of our loving relationships, despair can pull us under irretrievably. Holding hope for someone who's lost it can save their life.

The same relationship is often the source of both deep comfort and profound pain. Relationships are inescapably complex; two people who bring all of their wounds, biases, and defenses must manage somehow to find connection and understanding. How can we experience more harmony and intimacy in our relationships? The path to the best of what relationships offer will be familiar, building on everything you've learned so far about the Think Act Be approach. The heart of strong relationships is simple and obvious: being present with another person.

Be: Share Presence

Being fully present is the ground of our relationships, like soil for a seed. More than any words or specific actions, an open and embracing presence builds connection—and enriches the *think* and *act* practices we'll explore later on. We communicate that we're fully present simply by paying attention.

OFFER ATTENTION

I see my kids many times each day, but recently I had an experience of truly *seeing* them. I noticed the variations of color in their eyes, the characteristic way each of them moves, the idiosyncrasies that make them who they are. The warmth and love I felt for them in that moment broke my heart.

We all know the difference between being in the same room with someone and really *being* with them. When we're fully present, there's an unspoken acknowledgment that we are here together— that we see one another. Greater connection is always available when we deliberately step into the experience of being present.

We welcome connection with our children by giving them our full attention, if even for a few moments. We can find it with our friends when we orient fully toward them and see beyond our ordinary seeing—when we look, and then look again with our beginner's mind. We can even find it with people we've just met when we take the time to see them. I found it at the grocery store checkout, for example, when I asked the person in front of me what they were going to do with all those tomatoes. Suddenly we were in each other's worlds, sharing an understanding that stripped away the illusion of "stranger."

We don't have to add anything extra or overtly "spiritual" to find deep connection with another person. Spirit emerges in our relationships through our everyday words and actions when our attention is in the present. We go beyond minds and bodies by being fully present with our own body and mind. When we completely inhabit the shared here and now of this world, we find connection that transcends here and now, beyond space and time.

PRACTICE ACCEPTANCE

Being fully present with another person includes accepting who they are. It's easy to accept our favorite parts of someone, of course, and much harder to accept the ways they challenge us. I have no problem accepting my children's cheerful compliance, but I struggle to accept their occasional acts of defiance. It's not just that I like it better when they're agreeable; it can actually feel as if the challenging times aren't supposed to happen.

As simple as it is to accept others for who they are, we often stumble over it. I found myself struggling for a long time to accept that a boss I worked for had a difficult personality. I knew it intellectually because I'd been warned by everyone who had worked for him. Nevertheless, a part of me kept believing I could say or do the right things that would make this person stop being challenging.

I was nearly at my wits' end one day as I kept asking myself, "Why is he being so difficult?" Finally, it dawned on me: *because he has a difficult personality.* The long list of employees who had fled his organization testified to that truth. There was such relief in that acceptance. I no longer had to fight against reality or try to find the perfect way of responding to my boss that would turn him into a reasonable person.

It's especially important to be clear about what we mean by "acceptance" in the context of relationships. Here, we mean *acknowledging that a person is the way they are.* It doesn't mean we allow them to abuse us or that we invite toxic behavior into our lives. On the contrary, accepting that my boss was difficult was part of what led me to find another job. True acceptance leads to appropriate action and helps to free us from harmful relationships.

Have you found yourself struggling to accept a difficult relationship in your own life? Imagine what it might be like to accept that the person is simply hard to deal with, without believing that they should or will change. How might this kind of acknowledgment affect how you interact with them?

No matter what we're experiencing with another person, emphasizing acceptance will lead to better outcomes. Accepting others saves us from unnecessary aggravation because we no longer have to bang our head against a wall by trying to force them to change. Being truly accepted is just as profound—to be known, flaws and all, and received just as we are. More often, we get one half without the other: we're accepted but not really known, as with acquaintances at work; or known but not really accepted, as with a rejecting parent or critical partner. Being both known and accepted can help us to drop our defensiveness and more fully embrace ourselves.

We can also practice acceptance in our moment-to-moment interactions. The key is simply to say yes to what's happening at every turn. That doesn't mean we accede to every request or unwisely let down our guard. But we notice when we're trying to bend reality to our will. Instead of trying to wish away a difficult conversation or force someone to see things from our perspective, we choose to stay open.

This form of acceptance requires that we release attachment to our goals for an interaction. A goal forces our minds into an evaluative mode as we compare our fantasy of what we want to happen with our actual experience. We can't be fully in the give-and-take of a relationship if we're wedded to the outcome we want.

It's difficult to practice acceptance when we expect people to comply with our requests. Many times I've rushed to insist that one of my children follow the rules, for example, without asking them why they don't want to. There has often been a good reason, such as when our youngest child, Faye, was not going to bed on time because she was finishing a card for my birthday. Being open to these unexpected turns leads to fewer ruptures in our relationships.

Finding acceptance can be extremely hard when our ego is pitted against someone else's. The good news is that we'll have endless opportunities to practice accepting that people don't always agree with us and that sometimes we'll be misunderstood. We can accept that we may be treated unfairly and that people can believe things that make no sense to us.

We can even accept that our relationships won't always be what we want them to be as we release our attachment to feeling close and connected or having a meaningful conversation. With mindful awareness, we can step back and observe our reactions to a strained relationship, rather than letting the conflict completely determine our thoughts and emotions. With acceptance, we'll enjoy better relationships with less internal and external strife.

The simplicity of mindful connection is a strong starting point for addressing the many obstacles to sharing presence. Mindful thinking and acting can provide additional leverage for removing these obstacles and finding our way to one another.

Think: Check Your Assumptions

I'm grateful that I discovered cognitive behavioral therapy (CBT) relatively early in my marriage, over twenty years ago when I was in my master's program. One of the first things that drew me to it was realizing how my thoughts toward Marcia could affect our relationship. One evening not long after I'd learned the rudiments of cognitive therapy, she and I were arguing as I was preparing a frozen pizza for dinner (yet I have no memory of what the conflict was about).

In the middle of our fight, I realized that I was treating my assumptions about Marcia's intentions as facts. The way I was seeing her and the way I was feeling about her were based entirely on my beliefs, which might have been untrue. It hit me in that moment that this person looked completely different to me on the basis of what I was thinking about her. Negative thoughts made me see her in the worst possible light: unloving, unreasonable, unfair. When my thoughts were more positive, she seemed generous, warm, and kind.

The filter of my thoughts affected my feelings about Marcia, and these thoughts and feelings colored the quality of our interactions. Positive thoughts led to good feelings and harmonious interactions; negative thoughts led to bad feelings and conflict. Both could spiral toward either connection and intimacy or anger and resentment.

When someone I'm treating is struggling in one of their close relationships, we often examine their assumptions together. One time, it was my fourteen-year-old patient's assertion that her stay-at-home mother of five "did nothing all day"; it wasn't hard to discover that in fact she worked longer hours than anyone in the family. Another time, it was questioning whether a man's partner who left her things on the floor really expected him to pick them up. In reality, she was just unaware of her messiness.

The next time you're upset with a friend or family member, write down your thoughts about them. Then examine each one more closely. Is it 100 percent true? Does it tell the full story? Notice how these thoughts affect your feelings toward your loved one. Finally, think of at least one alternative way of thinking about the situation that might be more accurate.

The assumptions we make in relationships are often a form of *cognitive distortion*, or thinking error. With *overgeneralization*, we might think our partner is "always" critical or "never" likes the gifts we give them. Our actions and emotions that follow will be based on a mistaken way of seeing things, since the truth typically has more shades of gray.

With *catastrophizing*, we think a friend will never forgive us for a simple mistake—one we would easily overlook if the roles were reversed.

With *personalization*, we assume an aggressive driver was trying to stick it to us, even though they might have just been in a hurry.

Emotional reasoning could lead us to interpret the world through the filter of jealousy and falsely believe that our spouse is being unfaithful. As I realized that evening when I was preparing pizza with Marcia, my emotions often colored my beliefs about her in ways that weren't necessarily true.

And with a *false sense of responsibility*, we think our children's happiness is entirely up to us. *Think* helps us notice and question each of these faulty assumptions.

One of the mind's most common falsehoods is to direct "shoulds" at people who aren't doing what we want them to.

BEWARE OF SHOULDING

They should be nicer to me.
They should respect me.
They should admit I'm right.

I found myself shoulding at people in the Washington, DC, subway system years ago. I had no patience for the leisurely pace of tourists when I was rushing to catch a train for work or class. "They should get out of my way," I would think as I gritted my teeth. When I wasn't in a hurry, I would "should" at those who rushed past me: they should chill out; where's the fire? Whatever mode I was in at the time was the standard, and the rest of the world needed to obey my wishes.

In CBT, we treat shoulding as a thinking error because it's out of step with reality. When we say "should," we're suggesting that we have access to the Law of the Universe and that someone is violating it. But how could my preference dictate what the people around me ought to do? What was true was my *desire*: I *wished* the tourists would go faster or stand to the right on the escalator. I *wished* the harried commuters wouldn't make me feel that I was in the way. I could release unnecessary strain by accepting that my wishes won't always be granted and that no one was breaking the rules.

Shoulds also aren't very effective for getting others to change their behavior. The moralistic overtone of a "should" activates people's defenses and resistance, making them less likely to comply. It's also easy to argue with a should. "You should take out the trash" is debatable—the other person might argue that they *shouldn't* take out the trash. In contrast, "It would mean a lot to me if you took out the trash" is harder to debate. They might still decline our request, but they probably won't insist that it wouldn't mean a lot to us.

Pay attention for times when you're shoulding about someone in your life. How does your "should" affect your feelings and actions toward them? Is there a more helpful way to reframe your "should" statement?

We can also question our shoulds about difficulties in our relationships. Some of the worst fights are the ones we think *shouldn't be happening*. In reality, relationships are sometimes challenging, and we're bound to experience disappointment and conflict—even if we think it's ridiculous that we're arguing about something so petty. Rather than fighting the fact of these struggles, we can direct our energy toward accepting that sometimes we have petty arguments and then moving through them with as much grace as possible.

QUESTION MIND READING

"She thinks I look pathetic," I thought one evening as I waved at Marcia through our kitchen window. I was feeling weak and depleted as I walked back from taking the garbage to the curb, and I assumed I looked the way I felt. Automatic assumptions like mine are common in relationships, and we often don't realize we're making them. The mind is good at hiding its fabrications in plain sight, especially when they *feel* so true, as mine did.

However, I'd gotten in the habit of questioning my assumptions as part of my self-led CBT, so on this occasion my mind caught itself: "Does my wife think I look pathetic?" I had the perfect opportunity to test the accuracy of my belief: I could ask her whether she'd thought I looked pathetic when she saw me through the window.

Marcia gave me a look as if to say I was being ridiculous. "No, Seth," she said. "That wouldn't even have occurred to me." We can never be 100 percent certain that someone is being honest with us, but I believed her. My assumptions about what she was thinking had nothing to do with her and everything to do with how I saw myself. When I challenged my belief and found it was mistaken, I freed myself from the unnecessary weight of the criticism I had imagined.

When we think we're reading someone else's mind, we're often reading our own and projecting it onto them. Even if we can't ask someone what went through their mind, we can challenge our negative automatic thoughts about what they're thinking. At times, we may be right—spouses sometimes do think their partners are pathetic—but often we will be mistaken and will needlessly suffer.

I can't count the number of times I've had to catch my mind telling familiar lies that could alienate me from Marcia, such as "She doesn't care about me" or "She thinks I'm an idiot." There are a lot fewer problems in our relationship when I'm not assuming what her motivations are, what she's thinking, or what she "should" be doing. I can move beyond the mind's lies and half-truths that muddle my thinking and muddy the waters of our relationship.

However, the fullest expression of Think Act Be doesn't rest on convincing ourselves that people are kind or are thinking nice things about us. Mindfulness expands the range of cognitive therapy, as we've explored in earlier chapters, offering equanimity even when our negative assumptions are true or our relationships aren't going well.

RECLAIM HAPPINESS

Although it's hard to overstate the importance of our relationships, we don't have to outsource our happiness by giving others the ultimate power over our well-being. No one else is responsible for our happiness. We don't have to allow an aggressive driver to ruin our morning—why take on what isn't ours? We can let others' problems be their own. We need not allow others' false beliefs about us to upset our equilibrium; the contents of their minds can't change what we know is true about ourselves.

We can also hold others' beliefs more lightly when they disagree with us. Many of the conflicts we experience come from our unexamined belief that we can't be okay if people believe certain things, such as when they disagree with us about politics or about who caused a fight. But our peace of mind doesn't have to rest on getting someone to see things our way. With greater awareness, we can recognize when we're handing our happiness over to someone else, and we can choose to reclaim it.

> **Experiment with guarding your peace of mind regardless of what other people do. For example, resist sacrificing your well-being when someone says something you don't like. Question any automatic thoughts that say you can't be okay based on someone else's actions or words.**

Thinking errors don't just change how we see the other person; they can preclude loving, spiritual connection. Tending to our minds is therefore spiritual work because it moves us closer to what is true, and the seeds of love in our relationships grow in the light of truth. Cognitive skills won't iron out every wrinkle, even if frictionless relationships were possible. But understanding how the mind works, especially in the middle of a conflict, can be a saving grace.

We don't love others just with our heads, of course. We also have to use our hands. Mindful awareness and more skillful thinking can lead to actions that strengthen our connections to others.

Act: Live Your Love

The keys to connection aren't complicated. We simply bring our whole selves to the interaction, orienting our attention toward the other person with the intention of being completely present. Mindful presence bonds us together.

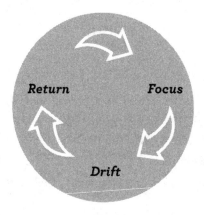

Figure 6

As simple as it is to communicate our presence, many things can get in the way. External distractions are easier to remove, as when we put away our phones or reading material and look directly at the person. But internal distractions can still grab our attention: "What should I make for dinner?" "I still need to respond to that email." "What if I'm getting sick?"

Mindful communication is meditation in action. We set an intention to focus on what's passing between the other person and ourselves; our attention drifts elsewhere; and we notice that our attention has wandered and reestablish our focus on the interaction. And the cycle continues (see figure 6).

In any interaction, we can be present both with the other person and with ourselves. We can pay attention to what our spouse is saying, for example, and notice their facial expression and body language. We can be aware of our own emotional reactions to the conversation as well—do we feel nervous? Angry? At ease? How is our body responding? We can aim to continue to open to what is happening, even when we have the urge to close down.

TUNE IN

I was fortunate to see many excellent doctors during my illness. My consultation with an endocrinologist stands out. Dr. Violeta Popii entered the examining room, sat down, and said, "Tell me your story." Then she listened intently, inviting me to share as much as I wanted to. Even though no clinical explanations for my symptoms emerged from my subsequent exam and bloodwork, I felt better after the visit simply from being heard.

It's a healing experience to be truly heard by someone. Listening and speaking to understand requires us to bring our whole being to the endeavor. We bring our mind, open to hear what the person is saying. We bring our body—the raw materials of our communication. And we bring our openhearted presence.

In *Missing Each Other*, Edward Brodkin and Ashley Pallathra offer the Dalai Lama as the embodiment of true attending: "In every single social interaction, he engages with people by giving them his complete and total attention. . . . Entering into conversation with him is a unique experience in which you feel completely seen and heard."[1] Offering another person our full attention sends a powerful implicit message, regardless of the words we say: *you matter*. This quality of presence is possible regardless of how much time we're spending with a person—even someone as busy as the Dalai Lama can bring all of himself to each brief encounter.

Much of the power of psychotherapy comes from having a dedicated block of time, free of interruptions and distractions, for tuning in. I feel truly in the other person's experience during the most fruitful therapy sessions, such as the one I described with Tina. Patients reveal their inner world to me, with words and beyond words, and I respond from a place I have access to only in that intense connection. We're completely in sync in an easy give-and-take, almost as if we're both under a spell. The same attuned presence is possible in any relationship.

> **It's easy to stop seeing the people you know well. The next time you sit with someone you care about, really take them in. Notice their eyes, their hair, and the way they move and speak. You don't have to try to feel anything specific or profound. Just observe.**[2]

With mindful presence, we can communicate that we're not just physically present with another person; we are available to them.

BE AVAILABLE

Unfortunately, I was often sending the opposite message when I first started working from home after shifting to part-time clinical work and closing my office. In the first months after giving up my office, I worked mostly in the common areas of our house when I wasn't in a video session with a patient. If my family members asked me a question, I would express my unavailability by the way I answered, usually by being as brief as possible and not looking up. I was present but not available. It wasn't long before I realized that being physically present but mentally and emotionally absent amplified a feeling of disconnection for those around me.

It's especially hard not to feel rejected when we experience distancing in person, even if the lack of attention isn't personal. Unavailable presence is one of the main pitfalls with our smartphones;

attending to the screen while around other people inadvertently sends the message that they're less important than our phones. They might know rationally that we aren't deliberately snubbing them, but on an unconscious level it doesn't feel good to be ignored.

Ideally, our physical presence and availability are aligned. When we're not receptive to being interrupted, we minimize the possibility of interruptions, such as by closing the door of our work space. I stopped working in the dining and living rooms to decrease the number of times my wife and kids would directly experience my unavailability. And when we're physically present, we aim to stay open to those around us. If we *are* interrupted at inconvenient moments, like in the middle of making dinner, we can express our unavailability with as much compassion and connection as possible (a work in progress for me).

It's a relatively rare experience to feel that someone is truly available, so your efforts in this direction will likely be noticed and appreciated. And as with any gift we offer, our attention is as much a gift to ourselves. Undivided presence is inherently rewarding, and it is doubly so when someone else benefits from it.

When you're feeling preoccupied with your own problems, see if it's possible to redirect your attention toward someone you care about who needs love or encouragement. Start by coming back to your center with a slow, grounding breath. Then ask yourself what the other person might need—a phone call, a quick foot rub, a jar of soup. Finally, show them support by offering what you can, which might help you feel better, too. Even if it doesn't, it will remind you that you can be of service even when you're not feeling your best.[3]

Being present goes beyond simple care and connection. We love what we pay attention to. To be present is to love, and love is the truest manifestation of who we are. The full expression of our relationships is more than a meeting of minds or having our bodies in the same room. It's a communion of spirits.

BE YOUR TRUEST SELF

True connection is inherently healing, and it meant everything to me when I was at the lowest points of my sickness. When doctors and other providers took the time to attend to my concerns, I didn't just feel heard. I felt loved. Whether or not we use the word "love," that's what we're looking for from others, especially when we're hurting. We don't want strictly technical support; we want to believe that the other person actually cares about us and that they'll be happier when our life is better.

Expressing love doesn't require finding the right words or any words at all. Some evenings when I was feeling hopeless, Marcia would play the piano and I would listen while lying on my back on the living room rug, feeling the vibrations of the music through the floor. The sound waves carried an energy that was beyond what my ears or body could detect. I felt love in every note as my tears released pent-up frustration and sadness.

Even in the midst of my physical and emotional struggles, those moments of profound connection felt exactly right—and for good reason. Our truest nature is to share the bond of love and connection. The spirits within us have an inherent affinity, since they arise from the same source and resonate with those of others.

When we recognize ourselves in another, our individual wins and losses no longer feel so personal. We can see that we're all living out more or less the same drama. Love inevitably emerges from connection—not a romantic love that craves attachment but a spiritual love that knows we're already joined. Ross Gay captured that fundamental ground of love and connection in his *Book of Delights*: "Caretaking is our default mode," he wrote, "and it's always a lie that convinces us to act or believe otherwise. Always."[4]

That lie is apparent in the grip of ego that separates us from someone who disagrees with us. It's hard to sense our common source and sustenance when we're at odds with someone. You'll probably notice a clamping down around your chosen beliefs and a strong urge to defend what you think is true. But ironically, in our efforts to guard our truth, we often move away from the deeper Truth of who we are—even if what we're saying is factually correct. As Jared Byas wrote in his excellent book *Love Matters More*, "Truth without love isn't true."[5]

We're most fully ourselves when we're giving and receiving love. We see this love in the constant caretaking that Ross Gay observed even among strangers: letting another driver go ahead of us, giving up our seat on the bus so a parent and child can sit together, even donating a kidney to someone we don't know. We extend ourselves over and over for people we'll probably never meet again.

Our highest calling is to strengthen our soul-nourishing connections. Few things were more important in my recovery from depression than rebuilding my relationships. For more than two years, I'd avoided people as much as possible because of my exhaustion and trouble with speaking. As I started to feel better, I knew I needed to reconnect with people on a regular basis, so I started meeting friends regularly for lunch and talking with my parents again on the phone.

I appreciated these relationships in a way I hadn't before and found myself sharing more openly about my struggles. My illness seemed to have stripped away the pretense of invulnerability and allowed me to connect with people more honestly.

True connection with our spiritual siblings is habit-forming. We're reminded how good it is to prioritize love, and the connection we enjoy compels us to continue nurturing our relationships. Like anger and fear, love is self-propagating. Love leads to love. And in contrast to destructive emotions, it builds instead of destroying.

No wonder I fell in love with CBT. Ultimately, it's a set of techniques that help us move toward love. I didn't know that explicitly when I was first learning it, but I sensed it intuitively. From the earliest years of my CBT training, I saw that effective treatment led to better relationships and strengthened the bonds of love. In the decades since then, I've also witnessed again and again the love that fuels each patient's work—love for kids, love for a partner, love of parents.

I continue to find benefit in my own relationships from the head, hands, and heart practices of Think Act Be. These tools help us to live in truth and to grow in love—and to express our love through meaningful work, the topic of the next chapter.

WORK IN
ALIGNMENT

When I was sick and depressed, everywhere I turned I saw things that needed my attention: the malfunctioning garage door, the broken fence, the pile of laundry in the hamper, the damaged caulk in the kitchen. "I'll do it later," I told myself each time I walked past an undone task. It didn't feel good to be neglecting my responsibilities, but I kept putting them off. Low energy and motivation made it hard to get started, and the inactivity strengthened my depression.

Life presents us with a continuous stream of responsibilities: doing our job, completing chores, caring for loved ones, volunteering. When our efforts are well aligned with these demands, work feels right. Kahlil Gibran wrote in *The Prophet*, "When you work you fulfil a part of earth's furthest dream, assigned to you when that dream was born."[1] Acting on the world in meaningful ways satisfies our fundamental need for *competence*: using our skills and abilities effectively to make a difference in the world.[2] Fulfilling this need is associated with greater energy and enthusiasm, higher life satisfaction,[3] and less depression.[4]

However, we often struggle to do the work before us. We might be avoiding tasks around the house or struggling to start our job search. Procrastination becomes its own source of stress as our work piles up. Or perhaps we're taking care of things, but it feels like a constant strain. How can we do our best work with less effort and more ease—especially the things we keep putting off?

This chapter describes how to use the Think Act Be approach to align with what needs to be done. You won't be surprised to hear that working in alignment begins with mindful awareness as we listen for what's required of us in each moment. We'll then explore how your mind can get in the way of getting things done and how to shift your thoughts and beliefs to make it easier to take care of things. The final part of the chapter presents powerful behavioral approaches that offer the leverage you need to fulfill your responsibilities. These practices will help you to embrace your tasks and to know the peace that comes from attending to them right away.

Be: Engage Mindfully

I once treated an emergency medical technician who wanted to find more peace at work through mindfulness. Greg understandably found it hard to stay present during the difficult calls he got on every shift as his mind imagined things that could go wrong. The resulting anxiety and stress often led to angry outbursts at his fellow EMTs. At first, Greg was concerned that mindfulness would interfere with his work, as if it were one more thing to remember to do. But as he discovered, mindful awareness sharpened his focus and made him *better* at his job, not worse.

ATTEND

Rather than dwelling on what might go wrong, Greg simply attended to what he was doing. No matter how many things he had to do as he cared for his patients, he could focus on one at a time and release his fearful fantasies of failure. "Some nights I feel like Neo from *The Matrix*," he told me, referencing a scene near the end of the movie when Neo realizes he's "The One." "All these things are coming at me and I'm working hard to handle them, but there's an effortlessness to it."

Working with greater presence changes our relationship with the task we're doing. We can let go of our attachment to the outcome and immerse ourselves in the process. The raging waters flow around us while we keep our single-pointed focus. Bringing our full attention to the work also provides clarity, helping us see what needs to be done. Sometimes we'll find that the work we had planned isn't the work we need to do. Picking up a sick child from school wasn't on the day's to-do list, for example, and threw a wrench in our schedule. With our awareness in the present, we can redirect our time as needed instead of rigidly insisting that we stick to our plan.

Paying full attention to our tasks also makes them more engaging. Something as mundane as brushing your teeth can be captivating when you pay full attention to it. You might try it the next time you have to do something you don't tend to enjoy, like taking out the trash or cleaning the bathroom. Just notice what's actually happening as you do it—what you see, the sounds you hear, the feeling of your body as it moves. Imagine it's the first time you've ever done this task. You'll probably find that it's not nearly as unpleasant as you expected.

> **Choose one task to do with full awareness, like getting dressed or wiping down the dining room table. Notice the feeling in your hands as you work, the sounds that your movements are making, and any other sensory experiences. When you find that your mind has wandered and you're lost in other thoughts, gently return your attention to the task without criticizing your mind for doing what it naturally does.[5]**

Opening to reality will also help us know when it's time for a change in our work. During the first couple of years of my illness, I kept slogging through full days of seeing patients. Mindful awareness helped me to drop the blinders of denial and see that I no longer had the physical and emotional stamina for full-time clinical work. Finally, I admitted to myself that I couldn't go on like that.

Awareness was necessary but not sufficient for creating change; even though I knew I couldn't keep working as I was, I refused for a long time to make a change. Mindful awareness had to be met with acceptance.

ACCEPT

We tend to tighten up as we anticipate a difficult change or an unpleasant task. But treating discomfort as the enemy can prevent us from doing things that are important and necessary, as when I needed to remove a three-inch cockroach from our bathtub years ago. My initial disgusted reaction to this creature was a firm *no*—this shouldn't be happening and shouldn't be my problem. When I accepted that this insect wasn't going anywhere without my assistance, I could focus on the job of corralling it into a jar and putting it outside.

Acceptance is the key to moving through disgust, tedium, fear, or any other discomfort that could get in the way of our work. With mindful acceptance, we can embrace the uncertainty of change, such as transitioning to full-time parenting or starting a new job. I finally reduced my clinical hours when I accepted that I had to do so for my health. Greg found that with greater acceptance, he could roll with the inevitable surprises he encountered in his job. Mindful presence helps us to be flexible rather than fixated on making things turn out the way we have in mind.

Being willing to move through discomfort can be a welcome surrender, as when I open to the cold as I wade into the icy waters off the coast of Maine. Rather than fighting against the freezing sensations, I simply notice them as an experience without labeling it "good" or "bad." I used this same approach for doing the tasks I was avoiding, like putting in a load of laundry, and accepted that I wasn't going to feel like doing it. When we're willing to be uncomfortable, very few things can get in our way.

Mindful awareness helps to clear away the many obstacles to doing our work so that we can align our efforts with our tasks. It can also help us to see when the story we're telling ourselves is making it hard to get things done.

Think: Engage Your Mind

"I just don't want my kids to feel like they *have* to work," Daniel told me in our eighth session together. We had been focusing, not coincidentally, on the stress that came from his own work as he built a successful internet advertising start-up. Daniel was on track to ensure that his children would have every financial advantage.

EMBRACE COMMITMENTS

There was a curious irony about Daniel's goal to spare his kids the burden of compulsory employment: he was miserable when he was idle for long periods of time. Daniel had actually retired in his early thirties with a plan to live simply on his equity share from the sale of the company he'd worked for. But after six months out of the workforce, he was restless and depressed. Research suggests a point of diminishing returns after just eight days of vacation.[6]

Nevertheless, work can feel like an unwelcome burden. Many of us believe that not having to work at all would be an ideal setup. "Always you have been told that work is a curse and labour a misfortune," wrote Gibran.[7] We might know on an intellectual level that work is rewarding, in the same way that we know exercise is good for us. But we also imagine how nice it would be to have no commitments and to spend each day doing whatever we want.

Our fantasies notwithstanding, having unlimited free time is a real problem, as I've seen in many of the patients I've treated. Their situations were especially challenging if they had exactly what Daniel wanted for his kids: not needing to work in order to provide for themselves.

My patient Ari found himself in that position after he graduated from college. His parents were providing free room and board with no expectation that he work, but rather than feeling free, he was

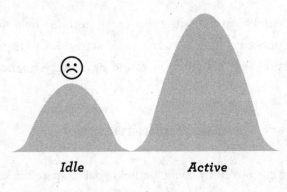

Figure 7

overwhelmed with options and paralyzed with anxiety. Without the motivation of bills to pay, he couldn't overcome the inertia that kept him from pursuing meaningful activities.

In order to get to a higher place, he would first have to go through a valley of challenges, such as the discomfort of changing his sleep schedule and the risk of failing (see figure 7). The longer he was inactive, the more bored and restless he became—and yet the valley he would have to go through became deeper as his anxiety grew and his confidence shrank. It was a cruel position to be in, with the short-term payoffs working directly against Ari's long-term well-being. The immediate and certain reward of avoidance overwhelmed the potential reward of applying to graduate school or finding a job. He wound up in a kind of limbo, feeling too idle to be at ease yet too comfortable to make a change.

For most of us, having too much choice in how we spend our days isn't helpful. It's not good to have to decide every day whether we'll care for family members or volunteer or go to our job. Our commitments take some choice off the table and become part of our identity, like becoming a vegetarian—we make an overarching decision that frees us from having to decide at every meal whether we'll eat meat. Commitments "bind us to the mast" (see chapter 6), making it easier to stay on course when we could otherwise drift into empty idleness.

> **If you've ever had an extended break, such as being between jobs or on summer vacation, think back to what you did and how you felt. Was it as enjoyable as you had expected? Did you do a lot of the things you'd planned, or did you find yourself languishing? How did it compare with times when you had more commitments?**

Most of us don't have the option to not work, and thank goodness. Having obligations almost always beats the alternative. I'm thankful that my illness and depression weren't so severe that I had to stop working completely. As challenging as work was, it also offered a break from self-focus, worry, and despair.

When your mind tells you that work is a curse, question the story. You can also be mindful of thoughts that lead to procrastination.

CHALLENGE PROCRASTINATION THOUGHTS

When I was tempted to put off tasks I thought would be a drag, like doing the laundry, my mind often said, "I'll do that later, when I feel like it." But in truth, I usually felt even *less* motivated later on. Each time I procrastinated, I experienced the relief of avoiding something I thought would be unpleasant. That relief led to negative reinforcement, which meant the behavior that caused the relief—in this case, avoiding the laundry—was more likely in the future.

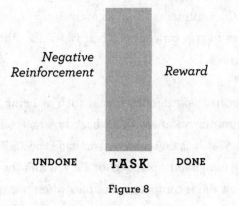

Figure 8

We also tend to underestimate the reward we'll experience from completing a task and no longer having it hanging over us. I was constantly surprised by how good it felt to empty the dehumidifier or replace a light bulb. When I was on the undone side of a task (see figure 8), the expected unpleasantness loomed so large that it was hard to see the payoff on the other side.

When you're tempted to put off something you need to do, look for the thoughts that could enable procrastination, such as "I can always do it later." Ask yourself whether there's a more helpful way of thinking about it—for example, "It will feel really good once this is finally done."

Recognizing the tricks of the mind made it easier for me to take care of my responsibilities. It helped to remind myself that my motivation wasn't likely to increase with more time and avoidance and that I would find more enjoyment than I imagined from having done something I was tempted to avoid. As I shifted my way of thinking, I completed more tasks and experienced the associated rewards.

With mindful acceptance, we can dig even deeper into our beliefs about tasks and emotions. Beneath the belief "I'll do it when I feel like it" is a more fundamental assumption: that we must avoid discomfort whenever possible. But we don't have to be at the mercy of our emotions. Saying yes to our experience includes accepting that we can do things even when we don't feel like it. Rather than ask what we feel like doing, we can ask what needs to be done.[8]

This is not to say that we should ignore our feelings, which can provide useful guidance. If we're miserable every time we go to work, for example, then it might be time for a new job. But for work that clearly is ours to do and needs to be done now, we can free ourselves from the trap of emotion-based avoidance. Most of the time, we'll find that the dread we felt was worse than actually doing the task.

RELEASE FEAR

The other big driver of procrastination is fearing that we'll do a bad job. Failing at a task is a punishing experience, which understandably we want to avoid. Many of the tasks I was avoiding when I was depressed were things I wasn't sure how to do, such as replacing the caulk in the kitchen. I didn't know which kind of caulk I should buy or how I should remove the old caulk, and I worried I would somehow mess it up. As a result, I kept putting things off.

Awareness of our self-doubts allows us to question the stories that lead to avoidance. How likely is it that we'll actually fail? Most of the time, these fearful stories are worst-case scenarios that don't come true. When we expect that things will turn out all right, it's easier to take a chance and get to work, as I found with the caulking job and so many other things I'd avoided.

Once again, mindfulness helps us to take the cognitive approach a step further and question the premise behind our fear. *Why must we avoid the possibility of failure?* Maybe we can open to uncertainty and accept that the outcome of our efforts is unknown. Greg, the EMT, learned to embrace the possibility of falling from his "high-wire act" rather than telling himself, "I must not fall!" This openness reduced his stress and tension even on his most difficult calls.

Mindful awareness and useful thinking are the foundation for action, the final dimension of Think Act Be.

Act: Take Care of Things

Taking care of things in the moment, without delay, isn't just about productivity or stress management. Meeting our responsibilities shapes our self-concept.

TRAIN YOUR BRAIN

When I was feeling really down, I felt indicted every time I saw an undone task around our house, like the blinking light on our base-ment dehumidifier signaling that it needed to be emptied. Avoiding these tasks eroded my sense of competence and reinforced my view of myself as a guy who doesn't take care of things. That self-image fed into my depression and further sapped what little self-respect I had.

It was important to question the belief that my self-worth was based on my productivity. However, *our brains are constantly making inferences about who we are based on observing what we do.* This is an underappreciated point that took me a long time to recognize. Putting off tasks sends our brains a subtle message about our abilities. Doing our work also sends our brains a message, strengthening our identity as someone who can take care of responsibilities.

With this realization, I stopped assuming it was "no big deal" to put things off. Part of my recovery from depression included taking care of my tasks right away. As many of my patients had found, simply doing what I needed to do was more rewarding than I'd expected. I didn't have to make any grand gestures and could work within my current limitations. Even the smallest accomplishment, like changing the filter on our heating system, was a significant moral victory. As I moved away from the fleeting reward of avoiding these tasks, I found the longer-lasting reward of seeing myself do them, which fed my motivation to keep crossing items off my list.

We often associate self-care with warm baths and taking time off, but it also includes tending to our tasks. Staying on top of things lifts our spirits when we're down and helps us feel effective when we're well.

BREAK IT DOWN

The first step in doing what we need to do is asking ourselves whether our task is the right size (see chapter 3). Tasks that are too big feel overwhelming and lower our motivation; if we don't know where to start or whether we have the stamina to complete a task, we're likely to avoid it. In contrast, rightsized tasks increase our motivation because it's easy to imagine ourselves completing them.

If we've been putting off doing laundry, for example, "catch up on laundry" may be too big and vague to spur us to action, especially if the dirty clothes have been piling up for a while. It's more helpful to define easily managed subtasks, like doing a single load of laundry per day. That way, it's abundantly clear what we need to do, and we're confident we can do it.

Rightsized tasks also include an obvious end point so we'll know when we're done. This is a crucial feature of effective task-related goals. We're often inclined to leave the end point undefined and say we'll do "as much as we can"—especially if we've been procrastinating and want to make up for lost time. But it's hard to commit to an open-ended task when it feels as if we're setting ourselves up for endless work. By making a well-defined end point—like "one load of laundry"—we keep our task manageable and can rest easy afterward, knowing we've reached our goal.

Pick an area of your home that you've been needing to declutter and organize, like a storage closet or kitchen drawer. Set aside ten minutes per day to work on it until it's done. If you'd like to start today but are feeling unmotivated, remember that motivation tends to follow action. See what happens if you go ahead and get started even if you don't entirely feel like doing it.[9]

JUST START

Breaking down our tasks facilitates the simplest behavioral principle for getting things done: *just start*. Do whatever it takes to leave the starting block. It doesn't matter if you take off sprinting, walking, or crawling—find forward progress. Any step in the right direction counts. Once we start, our efforts create momentum that tends to be self-perpetuating.

I've often used this approach with patients who were struggling to complete an important project, like Sam, who needed to write a term paper to finish his degree. The first step was to email his professor to confirm the topic he had chosen, which was his homework after one of our first sessions. When Sam came back the next week not having written the email, I encouraged him to do it right away during our meeting.

Sam was mildly annoyed by this suggestion—judging from his eye roll—which may have felt like micromanagement, but there was nothing more helpful for his progress than taking that first step and setting things in motion. By initiating contact with the professor, Sam pushed through his anxiety about starting. Once the ball was rolling, Sam found it much easier to keep going as he exchanged negative reinforcement for the reward of completing his work.

Starting our tasks puts us on the right side of costs and rewards, as seen earlier in figure 8. We get the reward of completing something, which boosts our motivation; at the same time, the project we're working toward gets smaller as we chip away at it. A mountain of laundry minus one load, for example, is a smaller mountain, and each load makes the task less daunting.

Breaking things down and just starting can be especially helpful for tasks that are deeply important to us but that we don't *have* to do, like starting a side business or writing a novel. Without incentives like a boss's expectations or a paycheck, there's often little to counteract our inertia. We can minimize the "cost of admission" by going small and dividing the project into tiny pieces—and then committing to the first step. When getting started feels almost effortless, we know the steps are rightsized.

Think of a project you've been avoiding. What is the first step you need to take to complete it? Make the step feel very doable by going as small as necessary. If it's repainting a room, for example, the first step might be finding the paint and brushes. Complete the first step today, and then plan to do the next small step tomorrow. Keep going like this, one manageable step at a time, until the task is completed.[10]

One of the commitments we can make to ourselves is to avoid work that isn't ours to do. Sometimes saying yes to life requires saying no to activity that pulls us out of alignment.

SAY NO

Working in alignment is more than simply being busy; it's leaving aside anything that could interfere with our essential work. Many of us have no problem with being active and instead struggle with chronic busyness. We often crowd our lives with *one more thing* that isn't needed and that diverts time and energy from what really needs doing. With mindful awareness, we can align what we do with the time and energy we have and say no when we need to.

It can be especially hard to say no when we might disappoint someone or lower their opinion of us. In my efforts during my illness to be a "good parent," for example, I would often say yes to family activities that left me feeling exhausted. Later in the day, the exhaustion made it harder to be a present, patient parent.

The pastor Eugene H. Peterson described what happens when we don't say no: "By lazily abdicating the essential work of deciding and directing, establishing values and setting goals, other people do it for us. . . . But if I vainly crowd my day with conspicuous activity or let others fill my day with imperious demands, I don't have time to do my proper work, the work to which I have been called."[11] Thomas Merton also invoked "laziness" to describe extraneous work: "Set me free from the laziness that goes about disguised as activity," he wrote, "when activity is not required of me."[12]

It takes work to say no to things we could do, or that others expect of us, that would draw us away from more important things, including rest (see chapter 8). It's essential that we "align our calendar with our values," as a patient of mine once described it, and protect time in our schedule for our most important activities. Mindful acceptance helps us make peace with the possibility that people will be unhappy when we tell them no.

As we embody mindful presence, tend to our thoughts, and act with purpose, we open ourselves to the spiritual dimension of work.

STEP INTO THE FLOW

I couldn't be truly present when I was saying no to my work, which put me at odds with the reality of what needed to be done. As I began to complete more of my tasks, I was actively expressing acceptance by receiving and saying yes to them. That affirmation brought me into alignment with my drive for competence and my longing for meaningful engagement.

Doing my work also allowed me to participate in the natural cycle of giving and receiving. Every day, I was benefiting from countless gifts[13] that usually escaped my awareness—the bed I sleep in, the house that shelters me, the dishes I eat from, the clothes and shoes I wear. Each of these necessities is the result of work that someone else did. The efforts of others continuously sustain me.

The most natural response to receiving is to give, as we see in any thriving system. Rivers give and receive water. Healthy digestive systems receive nutrients and release waste. The ocean tides push and pull. Giving is essential to life. If we don't offer back the effort that's ours to give, we cut off the flow; it is like an inhale without an exhale. "This is our work," wrote ecologist Robin Wall Kimmerer in *Braiding Sweetgrass*, "to discover what we can give . . . to learn the nature of [our] own gifts and how to use them for good in the world."[14]

We feel most fully alive when we step into the flow of life's moment-by-moment demands and offer the world our gifts. Embracing our work in this way is inherently a spiritual act. "If I am supposed to hoe a garden or make a table," wrote Thomas Merton in *New Seeds of Contemplation*, "then I will be obeying God if I am true to the task I am performing."[15]

Sacred texts also urge us to treat our tasks as sacred. In the Bhagavad Gita, Krishna says to Arjuna, "Know what your duty is and do it without hesitation. . . . If you want to be truly free, perform all actions as worship."[16] In his letter to the church in Rome, Paul says to his readers: "I appeal to you therefore, brothers and sisters, by the mercies of God, to present your bodies as a living sacrifice, holy and acceptable to God, which is your spiritual worship."[17] Eugene Peterson's paraphrase of Paul's words in *The Message* reads, "Take your everyday, ordinary life—your sleeping, eating, going-to-work, and walking-around life—and place it before God as an offering."[18]

The practices of Think Act Be help us to fulfill our duty "without hesitation," thereby granting us true freedom. With mindfulness, we can stay present and open to what needs to be done. Cognitive techniques remove the mental barriers to action. Behavioral techniques make the work more focused and manageable.

"Work is love made visible," wrote Kahlil Gibran. "And when you work with love you bind yourself to yourself, and to one another, and to God."[19] Consistently engaging with the world is a deeply satisfying expression of love, allowing us to "keep pace with the earth and the soul of the earth. For to be idle is to become a stranger unto the seasons, and to step out of life's procession." The love we express through our efforts allows life, and loving, to continue.

As we use our minds and bodies to act on the world and take care of our responsibilities, we come to see that we're doing more than keeping a schedule, meeting deadlines, or doing our job. We're aligning with life and with our nature as creative beings that act in meaningful ways. This alignment is a fundamental part of fulfilling our purpose. But as we'll see in the chapter that follows, your purpose encompasses much more than your work.

LIVE WITH PURPOSE

Everything we've covered in the previous chapters, from finding rest to tending to our bodies and relationships, prepares us to express our purpose in life. As you'll see, profound peace comes with knowing that you're living out exactly what you were made for.

* * *

"If I don't have a purpose, why am I even here?" Michelle asked me. She had come for treatment two weeks after surviving a suicide attempt. Michelle's plans to become a pediatric neurosurgeon had fallen through when she sank into a deep depression during her sophomore year of college. As her grades plummeted, her dreams of medical school began to slip out of reach. Now she felt aimless and adrift, with no sense of purpose for her life.

Michelle's question was understandable. Purpose unites the moments of our lives into something meaningful, giving us a reason to get out of bed in the morning and offer our best to the world. Purpose makes our sleep sweeter as we rest in knowing that our lives matter. Without a clear purpose, life can feel hopeless. I recognized Michelle's despair in myself when my illness interfered with my work as a therapist.

How can we cultivate an overarching purpose that remains even if we lose a job, our health, or our abilities? As I discovered, I was looking for my purpose in the wrong places. With mindful awareness, we can never truly lose our purpose.

Chasing My Purpose

I knew I'd found my calling when I started my clinical practice. I could see the difference I was making in my patients' lives, had a beautiful office I could walk to, and loved being my own boss. I was offering help to those who were hurting while providing financially for my family. I had found what I was meant to do for the rest of my working years.

But that was not how it turned out. As my illness grew worse, I struggled to meet the demands of a busy practice. I no longer had the stamina I'd relied on to work long hours. Nevertheless, I kept dragging myself to my office in the early morning through the worst of my sickness and slogged my way through each day. "This is my calling," I insisted. "It's what I'm meant to do!"

By the time I stumbled home in the evening, I was exhausted and had little left for my family other than impatience and irritability. All the while, I thought I could get back to my normal routine of seeing twenty-five to thirty patients per week. I could relate to Michelle's hopelessness—how could I go on if I lost my function? Like a broken knife, I would belong in the trash. If I couldn't provide therapy and a paycheck, I felt pointless.

And yet I saw countless examples of people who had lost much more than I had and seemed to be living with an abiding sense of purpose. Maybe I was mistaking my *role* for my *purpose*. After all, there are many ways to offer cognitive behavioral therapy besides one-on-one therapy sessions: books, blogs, podcasts, apps. The delivery method wasn't as important as the goal of helping others to experience less suffering. Perhaps I was the chef instead of the knife. Even if I couldn't offer full-time therapy, I could still find ways to live out my calling.

What do you believe is your purpose? Consider how your sense of purpose may have evolved over time. Have you ever felt as if you couldn't find it?

But once again, my relationship with my purpose felt tenuous. I had to be healthy enough to keep working. I had to be successful. I insisted that I had to be able to do *this type of work*, or have *that role*, or earn a certain income. In order to meet those demands, I had to seek safety and avoid pain. For more than three years, I begged for relief from my suffering. I believed that healing meant that my health problems would disappear and I would be able to work as I had before.

Without realizing it, I was cultivating an ego-based sense of purpose, demanding that things go my way and resisting the possibility that they might not. I couldn't settle into the striving of the ego because nothing about my work was guaranteed. Basing my purpose on my work brought a feeling of unease and the constant threat that I could lose what I had.

When maintaining our sense of purpose depends on holding on to something or reaching what we're chasing, we are being controlled by the ego's attachments. Frustration or disappointment is inevitable when we ground our purpose in a job, or in being a caretaker, or in carrying out a creative endeavor. Athletes will face a crisis of identity when they can no longer play if they find their ultimate purpose in their sport. If parenting young kids is our ultimate purpose, doing it well will probably ensure that eventually we'll no longer be needed. If our career is our purpose, we'll feel lost in retirement.

We may find great meaning in these functions, but we need a purpose that can't be taken away. Otherwise, all hope is lost when we find ourselves in a situation like Michelle's. She saw no path to her purpose, and so her life seemed pointless—until she discovered a purpose that is deeper than our roles and more durable than our health.

BE: FIND PURPOSE IN PRESENCE

The king in Leo Tolstoy's short story "The Three Questions" set out to discover his purpose. By the end of the story, he realized that his purpose was based on presence in the now "because it is the only time when we have any power."[1] Our charge is to be present for those we're with and "to do that person good, because for that purpose alone [were we] sent into this life." Mindful presence is the foundation for purpose.

Brother David Steindl-Rast reached a similar conclusion in *Music of Silence*. The entire rationale for the monastic way of life is "most succinctly described as an effort to live in the now,"[2] he wrote, with the intention to respond to "a series of opportunities, of encounters."

Paradoxically, my fear of losing my purpose became self-fulfilling: as I clung to the work I thought I had to do, I sacrificed the purpose that was available to me. I couldn't be present for those around me when I was preoccupied with what I might lose. My fear and clinging eclipsed everything that really mattered.

Our purpose is to be present. There is peace in that simple recognition. Our true purpose cannot be taken away from us. It is never something we cannot do. We're not tools that can lose their purpose, like a broken knife that we discard. Nothing outside of us can interfere with what we're meant to do—to be present, no matter what's happening. Every single moment of our lives presents the opportunity to fulfill this purpose. Even this very moment.

When you know your real purpose, you find a sense of calm, as if recognizing an old friend: "Ah, there you are." It's not like becoming a brain surgeon, as Michelle realized, which rested on countless things beyond her control: grades, admissions decisions, her mental and physical health. You don't have to make things work out in a particular way so you can fulfill your purpose. There's no anxiety about your purpose when you can't lose it. You can rest in knowing you have everything you need, always. With every breath you are exactly where you need to be.

Through my illness, I discovered that being whole was not a matter of making everything just right so I could get back to my life's purpose. Healing became a process of aligning with my present situation so I could bring my best—whatever that was—to each day. This is a dependable kind of healing that we don't have to beg for because alignment is always available. It's as close as listening to our body and mind and offering what we have.

We can find new life when we decide to live now, just as we are. No matter where we are, how we're feeling, or what's happening, we can be present. To be present is to be available, and when we're available, good things follow. We can respond flexibly to whatever *presents* itself—whatever makes itself present. When we're centered in the now, we can attend to our own needs and those of others. We can work when it's time to work and rest when we need to. We can give and receive love.

Most of us tend to listen to others with one ear and attend to our thoughts and worries with the other. Choose one interaction today in which you'll bring your full presence to the other person. Pay close attention to them. Focus intently on the words they're saying, their body language, their eyes and their facial expression. Notice how you're more available to the other person when you purposefully attend to them.[3]

Presence—the *be* in Think Act Be—allows us to match our actions to the needs of the moment. As soon as we open into mindful presence, we're aligned with our purpose. "Do what you're doing," wrote David Steindl-Rast. "This loving response to the call of a given moment frees us from the treadmill of clock time and opens a door into the now."[4]

As always, being is the foundation for effective thinking and directed action.

THINK: KNOW YOUR PURPOSE

Some of the mind's most misleading tales are about our purpose. It's easy to fall into all-or-nothing thinking about our purpose and insist that we *have to* achieve a specific outcome, that the alternative is nothing—zero. Michelle's mind told her that she had to be a neurosurgeon or else she was nothing. She had believed this story for so long that she didn't recognize its black-or-white quality: it had to be *this* and *nothing else*. Sadly, she took the "nothing" so seriously that she became suicidal, believing that the only choices she had were reaching her goal or ending her life.

Black-or-white thinking also leads us to believe that our purpose is worthless if it's not something epic. Michelle didn't just want to save lives—she wanted to change the world through her medical work. In our sessions, she came to see that she was *discounting the positive* by ignoring the good things she brought to the world—first and foremost, her unique presence—whether or not she was a doctor. When Michelle expanded her beliefs about what her purpose might entail, she rediscovered her self-worth and no longer felt like ending her life.

Most of us equate purpose with *doing* and discount the value of our presence. This thinking error functions as a society-wide delusion, causing us to believe that our productivity determines our worth. I have certainly recognized this core belief in myself, and it takes persistent awareness to see through it.

Begin to challenge the premise that your purpose in life is to achieve. Distrust any mental story that insists you must do more, get more, attain more, or be more and that tells you these

are the paths to purpose. Anyone who has followed them will tell you that they do not lead to what you're hoping for.

There's nothing wrong with goals, and we can find real meaning in our work and relationships. But we need to be careful not to tie our well-being to things that pass away. Even if our dreams die, we don't have to die with them.

Mindful presence and right thinking together set us up for purpose-driven action.

ACT WITH PURPOSE

Our purpose is the same—to be present—regardless of how we choose to spend our time. However, this is not to say that it doesn't matter what we do—quite the opposite.

Come Alive

There's a world of difference between work that's meant for us and work that is not. In *Wishful Thinking*, Frederick Buechner describes finding "the place where your deep gladness and the world's deep hunger meet."[5] You come alive when you do "the kind of work (*a*) that you most need to do and (*b*) that the world most needs to have done."

Your calling lies at the intersection of your abilities and your interests—where what you're *able* to do matches what you *want* to do, which infuses your efforts with the full force of who you are. Working in this way will bring you more fully into the world, facilitating your presence in it. On the other hand, being fully present is more challenging if your work isn't right for you. It's hard to be available in the moment when your situation doesn't fit who you are; it's like wearing a shirt that's too small.

I felt the discomfort of a poor fit when I was pouring hours into research studies, knowing deep down that my work contributed little of real value to the world. I found no deep gladness in the work, and it wasn't meeting a deep need. I experienced another poor fit when I was forcing myself to work longer clinical hours than I could sustain. The mismatch between my tasks and my abilities was a continual source of strain. With mindful awareness, I've been able to move toward work that feels both life-giving and sustainable.

Mindful awareness helps us to notice when we feel most present and alive and to direct our energy accordingly. We can leave aside the things that drain the life out of us. We can choose work that helps us to be present. We can stop trying to shoehorn ourselves into work that we're not able to do; our calling will never be beyond our abilities.

What kind of work always leaves you feeling drained? Compare those tasks with ones that make you feel most alive. Are there ways to do more of the things that fit with your abilities and interests?

The work you do can *serve* your purpose but is not itself your ultimate purpose. That space is reserved for presence.

Make Space

If we value mindful presence, we need to create opportunities for it in our daily schedules. These opportunities might include meditation and other mindfulness practices, but even meditation can become one more thing to accomplish in the day. "It is no use trying to clear your mind of all material things at the moment of meditation," wrote Thomas Merton in *New Seeds of Contemplation*, "if you do nothing to cut down the pressure of work outside that time."[6]

Being too busy leads to more stress and less availability to meet the needs of each moment. Cramming our days full of activity leaves no room for being, like graphic design without white space or musical notes without silence. Planning space and silence into our days will make it easier to notice what's important. It will also be easier to act *purposefully*—with intention—when we move from a balanced center instead of flailing our way through the day.

Serve Your Purpose

Part of the gift of presence is that it opens the possibility of serving others. Mindful presence reveals what others need and how we can respond. Service flows naturally from being present for the people around us.

However, it's often hard to attend to others' needs, especially when we're struggling ourselves. We tend to turn inward when we're going through a hard time as we try to take care of ourselves and conserve energy. My attention was constantly on my own distress during the lowest points of my sickness and depression. I was fixated on how hard everything was and how I just wanted to feel better. Others' needs were lost in a sea of self-focus.

This experience brought home to me how deeply unsatisfying it is to be habitually preoccupied with our own concerns—and, ironically, how exhausting. Dwelling on our problems feeds a downward spiral of anxiety and inward focus and cuts us off from life-giving opportunities to consider others. Like many of my patients, I found that I could be present and extend myself for others even when I felt broken. It was enough to offer up what I had, even when it was an empty cup.

We don't have to make grand gestures; we might just ask how someone's day was or do a small favor for them that makes their life a little better. We can even use our own struggles as cues to think of others. If we're in pain, we can ask ourselves who else is hurting and what we might do to relieve their suffering. When we're feeling discouraged, we can think of someone who needs our encouragement.

None of these redirections are meant to ignore our own well-being. When we're mindfully present, we'll respond to our true needs; attending to the one we are with, as Tolstoy urged, includes self-care. Having done what we can for ourselves, we then release rumination and worries that aren't productive and direct our energy outward to those around us.

When you're dealing with worry or anxiety, ask yourself how you can show love to the people in your life. Look for an opportunity to meet the needs of someone else, perhaps in ways they aren't expecting. Focus on the act of loving rather than on waiting to feel loving. Love is often the antidote to fear.[7]

Caring for others is often the most caring response we can give ourselves, and it can be crucial for our own healing. Part of my recovery, for example, included sharing the responsibility of putting our kids to bed, which my wife had been doing on her own for many months with my frequent crashes in the evening. Making that commitment helped to pull me out of my own cares and offered many sweet moments as I read or sang to our kids at night. No matter where we find ourselves, we can align with our nature as loving, serving beings.

At the same time, it's important not to seize on being of service as our sole path to purpose. When I thought my purpose was to serve my patients and my family's financial needs, I was constantly running into apparent obstacles—illness, vacations, my wife's appointments that required me to be at home. These types of things may conflict at times with our work, but they can't interfere with our fundamental purpose.

When we're present with our whole being—mind, body, and spirit—everything we do is part of a unified whole.

INTEGRATE

When I thought my purpose was to advance the world of trauma treatment, I was irritated at having to calm a crying baby at night instead of working on my grant application. Helping my daughter get to sleep felt like an obstacle to what I needed to do. But as we align with our purpose as presence, we find that it's just as available to us in our "obstacles" as in what we're "supposed to be doing."

A clear awareness of your purpose provides coherence and integration across all the parts of your life. Every moment offers the opportunity to be present and serving your purpose. Our responsibilities at work and home aren't so much "balanced" as integrated because our singular purpose is always the same. True purpose leads to harmony instead of dissonance across the domains of our lives.

An integrated sense of purpose also fortifies our lives against the unpredictable, just as a building with architectural integrity holds together in an earthquake. When our lives are shaken, we don't fall apart. A family emergency or a flooded basement can scrap our plans for the day and we can know we're still serving our purpose. We can work for years building a career that crumbles for reasons outside our control and yet trust that our life is no less meaningful. These challenges are disappointing—even heartbreaking—but we can know on a deeper level that our purpose is never in doubt.

There's beauty in a unitary sense of purpose. Every facet of our lives works together as in a well-tended vegetable garden; whether we're planting or watering, raking or weeding, our activities are oriented toward the harvest. In the same way, every aspect of our lives—sleep, exercise, relationships, work—is connected to the others as part of a bigger purpose. Each activity offers the same opportunity: to be present, doing exactly what we're doing.

All the practices of Think Act Be ultimately allow us to fulfill our true purpose. Being receptive to the present facilitates healthy thoughts and behaviors. Right thinking and action in turn make room for presence. Our head, hands, and heart are united in a common purpose—and one that's everlasting.

UNENDING PURPOSE

Toward the end of her treatment, Michelle realized that her purpose was not something outside of herself. "I don't have to be more than I am," she told me. "I just need to be." Emphasizing presence provided clarity as she explored different directions for her career. She still felt a good deal of anxiety about her future, but she no longer felt desperate. Nothing could touch the core of her identity and purpose.

Our deepest purpose inheres in a state of being rather than in what we think or do. When we wake up to the spiritual dimension of our purpose, we find that it transcends time. Our deepest purpose is always available because it lies in the present, and to step into the present is to enter eternity.

You can sense this continuous connection deep within yourself when you fully inhabit a moment. When we're present, "everything has meaning, everything makes sense," wrote David Steindl-Rast in *Gratefulness, the Heart of Prayer*. "You are communicating with your full self, with all that is, with God."[8] All questions slip away, you're doing what you're doing, and it just feels right.

When we work, our purpose is to be alive to the work we're doing. When we're resting, we fulfill our purpose by being present in the pause. When our strength gives out and we approach the end of our days, we can be present to life itself and to each remaining breath. And when our time on Earth is over and we breathe our last, we are present to whatever begins when the breath ends.

Our purpose is fulfilled by fostering presence, in this moment as in every other. As you come alive to the present, you experience a complete and profound sense of belonging, right here and exactly as you are. In the core of your being, you know you're home. Coming home to ourselves completes the circle of Think Act Be, and it is the focus of the next and final chapter.

◇ 13 ◇

COME HOME

And when I thought that I was all alone
It was your voice I heard calling me back home.

—Rich Mullins, "Growing Young"[1]

This book began with listening for the inner voice that's calling to you. The culmination of hearing and following that voice is coming home to your Self—to the truest *you* that's a part of the divine. In this final chapter, you'll learn about the wholehearted love that awaits you when you come home to yourself, and the everyday miracles that become apparent when you know you're deeply loved. As you feel truly at home in yourself and your world, you'll discover ever greater connection with your inner voice, and the Think Act Be cycle continues.

* * *

One evening during my recovery from depression, I was hit by a realization of the nature of divine love. It happened while I was making dinner and listening to a concert by the Christian singer and songwriter Rich Mullins. Between songs, Mullins said he was grateful that God is like a proud parent who treats the messy scribbles of our lives as "the most beautiful art."[2]

In a flash, I understood that divine love is not a begrudging, half-hearted love or a detached tolerance, as I had long assumed. It is unconditional, wholehearted, and inescapable, like a parent's love for their child. That night I fell asleep feeling welcomed, embraced, and immersed in benevolence. It felt like I was home.

All of us long to come home. We long for a place where we can be fully ourselves, lay down our cares, and know that we're loved exactly as we are. And yet this experience of deep belonging so often eludes us. We feel uncomfortable around others and cut off from ourselves. We doubt our own worth and long for more than we have. How can we feel at home in our lives and at peace with ourselves?

Coming home is the culmination of the Think Act Be journey, as I discovered that night in the kitchen. All of our experience is calling us home: to one another, to this Earth, to this moment. Every chapter of this book has been about listening for that call. It's the call that brings a person to therapy and the call I heard on the couch when I felt lost and alone. We can hear it in sacred rest and in the ecstasy of physical movement. The same call spoke to me as I dived into the water of the Delaware Bay with the words from *Moana* ringing in my ears. It's in a leaf of lettuce and in the face of our loved ones.

How often do you experience a deep sense of love and belonging? How easy is it for you to love yourself? Consider what it might mean to feel at home.

The specific religious and spiritual experiences I'll describe are not a template for anyone to follow because the home I returned to wasn't a place or a set of beliefs. It was being at home in myself. This home is everywhere and always as a state of being. But even though the call is constant, we're often unaware of it, as I was for many years.

Hear the Homeward Call

I shut my ears to the call in my early twenties when I left the Christian fundamentalism I'd been raised in. I had tried so hard to believe in the miracles we were promised, praying fervently and fasting with my congregation as I asked God to heal the sick among us. All the while, young and old people died of cancer, my own illnesses and injuries seemed immune to prayer, and even the pastors who preached divine healing relied on medication for chronic conditions. Eventually I felt deceived when the promises didn't materialize and alienated by the emphasis on an angry, judgmental God who seemed constantly ready to strike me down.

It was painful and disorienting to lose my spiritual footing. For nearly two decades, I bitterly rejected anything religious or even spiritual. I derided Christianity, profanely mocking the idea of a loving God. I told myself I was a strict materialist as I tried to believe that all of human experience could be reduced to observable properties of matter. I wanted to leave no room for the supernatural, and yet I couldn't shake the sense that there was more to life than I acknowledged.

I was half aware of many gentle tugs at my sleeve during the years that followed. "Something is missing," a quiet and persistent voice said. Although the call to enter fully into our experiences is ever-present, it usually doesn't hit us over the head. I sensed it most often in dreams, songs, poetry, and natural beauty and in the stillness of mindful awareness.

I continued to be moved in spite of myself by long-familiar passages from the Bible, such as Psalm 42:1: "As the deer pants for streams of water, so my soul pants for you, my God." I couldn't explain away the swell of emotion I felt while singing with our choir in Bar Harbor, Maine: "O love that will not let me go / I rest my weary soul in thee." I cried during the movie *Junebug* when a trio sang "Softly and Tenderly" in a church basement that looked just like ones I'd known growing up: "Come home, come home, ye who are weary come home."

Many years before my illness, I heard a clear call late one night as I was sitting at my basement desk working on a research article. I paused from my work at one point to listen for the first time to a new song by Mumford & Sons called "Awake My Soul," from their enormously popular debut album *Sigh No More.* I wasn't expecting the powerful wave of longing and sadness I felt as I listened to the simple lyrics.

When the chorus began, I suddenly began to cry. It wasn't a respectable tear or two running down my cheeks that I quickly brushed away; I was weeping with my whole body, my face in my hands. I didn't understand my tears or my anguish, but I knew I had felt half dead inside for a long time. The pain of separation as I cried was undeniable—separation from reality, from others, from myself. I felt I'd been missing the most important parts of my life.

Even my dreams around that time were waking me up to a deeper understanding of what it means to be alive. Shortly after that night in the basement, I had variations of the same dream three nights in a row. In each one, I was dancing euphorically and swaying together with a group of people I didn't know as they sang in a language I didn't understand. It was strikingly similar to the many "altar calls" in the Pentecostal churches of my youth, when everyone was invited to the front of the sanctuary for an emotionally intense time of praying and singing together, which often included speaking in tongues.

"I feel like something is calling me," I said to my wife after the third dream, "but I don't know how to respond." The only template I had for spiritual practice was going to church, which didn't feel right. After our conversation, I decided to enroll in a yoga class, which is where I discovered mind-body practices, the power of the breath, and the peace that's found in presence.

During this same period, I had the death dream that I recounted earlier in this book, which showed me a level of connection that I didn't know was possible: union with all that exists across time, space, and even death. "How had I abandoned this?" I wondered. It felt as if I'd been plugging my ears through my twenties and thirties and saying I didn't believe in sound. And then I'd rediscovered my sense of hearing—words, laughter, music, birds singing.

As I continued to say yes to the call I was sensing, I discovered a tiny book on secular Buddhism that opened my eyes to the power of mindful awareness. It had been on my shelf for many years but hadn't caught my eye before. Nearly every page blew my mind with its incisive description of things I had half sensed but never put into words, such as the tenderness approaching sadness that infused all of life when I was really close to my experience. It was the sweetness that made me cry when I paid attention to a raisin or tasted life in a piece of lettuce from the garden.

I soon embraced secular Buddhism, with its emphasis on being present in our experience and embracing life just as it is, and I began to practice daily meditation. I learned through Buddhist writings of all-encompassing love and oneness and a universe that's not looking for an opportunity to punish us. I found peace and acceptance in secular Buddhism, which became my adoptive spiritual home. Maybe it was okay that I was flawed. Maybe I could accept myself just as I was.

Nevertheless, I still found myself searching for something I couldn't quite name.

FIND HOME WITHIN

Most of us seek security and belonging in a home outside ourselves. We look for it in relationships, in work, in acclaim or material things. In reality, the home we long for is always right here. It doesn't exist anywhere else. "This is a country whose center is everywhere," wrote Thomas Merton in *New Seeds of Contemplation*. "You do not find it by traveling but by standing still."[3]

But that doesn't stop us from searching for it everywhere except where it is, as if we were tearing the house apart trying to find the keys that we're holding in our hand. The answer must be *out there* somewhere, something more than just ourselves, because we're convinced we're not enough. This belief in our own insufficiency shows up in our core beliefs about ourselves and the world that drive our thoughts and actions. These deep-seated assumptions are often about being inadequate or unlovable: "You have been weighed on the scales and found wanting."[4]

Where do you find yourself looking for something that's missing—something that would complete you? Think about

the core beliefs about yourself that might underlie any sense of needing something more.

So much of our suffering comes from struggling to find *something more*: More than myself. More than the person sitting across from me. More than this house, this table, this meal, this spoon. More than this moment. We're always looking beyond our little lives, hoping to find something better. From this sense of deficit, we try to gain the whole world but risk losing our soul.[5]

This theme shows up in countless song lyrics, books, and movies—including the Disney movie *Frozen II*. (I'm required to watch Disney movies as a parent of young kids.) Elsa is pursuing a being she senses—someone or something she's known for as long as she remembers and yet has always eluded her. As she chases a voice and a light, she sings with greater insistence and begs the presence to reveal itself.

That hope and desire lives in all of us—for something that will satisfy our longing and complete us. We seem to get so close at times, as Elsa does, like remembering the feel of a dream but not being able to recall what it was actually about. Elsa stops at nothing in her desperate search, having ridden a magical horse that nearly drowned her before carrying her across the sea. As the pursuit continues with increased fervor, Elsa commands what she's chasing to reveal itself. She is certain that if she looks hard enough she can find what she's always been waiting for.

We make the same demand of the things we pursue—the people in our lives, our careers, every diversion that would take the place of what we really need. We think the solution to the void we feel lies in gaining something we lack. But what we're looking for has been with us the whole time, as Elsa discovered. The music builds to a crescendo and she finally finds her answer: *That she herself is the one she's been waiting for.*

The *you* that you have been waiting for is not the ego part of you that fortifies its separateness, like a drop of water that tries to own the ocean. What a disappointment that would be, if the "small *s*" self was the fulfillment of all that we long for! Instead, this is the you that's part of something greater. It's who you truly are. You don't have to compensate for a perceived shortcoming or falsely aggrandize your tiny presence in a vast cosmos. You can stop trying to own the world because you belong to the universe.

There is a fundamental truth of who you are that's deeper than your negative core beliefs. If you dig within yourself and strike what seems to be the bedrock of your inadequacy or unlovability, keep digging. You've not yet found your true core. In the deepest part of yourself is a divine knowing that you have everything you need. You've always been enough. You're okay exactly as you are. Your shoulders relax and the breath gets easier with this realization.

"Why abandon who you are?" asked the Dominican mystic Meister Eckhart. "Why not remain in your self and draw from the deep well of your own good? For each of us holds all the truth in ourselves."[6] As I discovered in my own wanderings, you can never really leave your true home. It's a part of you. Coming home means reconnecting with what has been there all along. Your homecoming is ready for you.

No matter how far we've wandered from the true home within ourselves, we don't have to retrace our steps to find our way back. Homecomings happen in a single moment, like a light switched on in a dark room. We may have sat in the dark for a long time, but suddenly we're aware of things we hadn't been able to see. Wave after wave of insights and implications washes over us as we realize that the light of our new understanding changes everything.

I witnessed many dramatic homecomings to the true self in my therapy office, as with Paul, whom we met in chapter 1. He had never felt good enough, and he had looked for completion through his mother's love, through alcohol, and through a series of romantic relationships that never quite satisfied. In the process, he had abandoned himself. Through his practice of mindful cognitive behavioral therapy, Paul came at last to rest in relationship with himself—with the five-year-old boy who'd been abandoned and alone and with his adult self who carried those wounds of long ago. I knew he had found his way back home when he told me he felt love toward himself; the call homeward is always a call to love.

When we're finally home in ourselves, love inevitably emerges. Nothing lets us know we're truly home like the full awareness of our lovability.

KNOW YOU ARE LOVED

I felt much more alive to my experience as I started practicing meditation and other forms of mindfulness. I realized how much I'd been missing in my relationship with my wife and kids, and I even saw strangers with a sense of recognition as I walked down the street. I found myself smiling irrepressibly at each person I saw and didn't care if I looked a bit ridiculous.

Buddhism even helped me to see Christianity through a new lens and to recognize deep truths in the things I'd been taught: that spiritual connection is essential, that love is foundational, that the eternal resides in each of us. Dogmatically literal interpretations of these teachings had obscured their fundamental message—one that harmonized with the insights I'd found in Buddhist principles. I was left with a completely different understanding of the religion I was raised in. Many of the words were the same—sin, faith, redemption, death, Christ—but the definitions had changed. I finally understood that God is love.

And yet the spirituality I encountered through Buddhism didn't get to my core. There was a disconnect between my explicit belief about the nature of God and what I felt to be true in my gut. It was the same dissonance I'd encountered so often in my therapy practice, when a person knew their self-hating thoughts were unrealistic but they still *felt* like the truth.

I still craved the love and approval of my estranged "birth God," the one I'd grown up with and that I'd fled when I ran away from Christianity. My deeper convictions still told me that God was vaguely irritated with me and itching to punish me. God's patience might be perfect, but I was sure I'd worn it thin.

The emotion that poured out of me when I listened to Rich Mullins's words had been building for decades. As I cried in my kitchen, the Apostle Paul's encouragement to the Romans held new meaning for me—nothing "in all creation will be able to separate us from the love of God."[7] At last I was home.

WITNESS MIRACLES

When I was growing up, I thought a miracle was something outside the realm of everyday events. It would be miraculous if terminal cancer disappeared or if a broken bone was suddenly healed. When these extraordinary signs failed to happen, I decided that life was nothing more than our mundane experiences—nothing magic, nothing divine.

But when I looked more deeply into my normal experience, I found that there *was* more than I realized—not more than this life but more *in* this life. My body could bind up cuts and heal my colds, with a power beyond my conscious awareness or control. I could offer the right conditions—cleaning a wound, getting enough sleep, eating nutritious food—and my body would do the rest.

As you go about your day, notice how your lungs enable every activity you do. Feel how your breath adapts to each demand: sharp and shallow when you're anxious, deeper after you climb stairs, slow and even as you drift toward sleep. You can also be aware of the automatic calming effect of deliberately slowing your breath.[8]

Our minds and hearts can heal, too, as I've witnessed in my therapy sessions and in my personal life. Just as with physical healing, we offer the right mental and behavioral conditions and our nervous system takes it from there: We approach our fears and they diminish. We find engagement and our depression lifts. We change our habits and our sleep improves. We find rest and our stress system calms down.

The routine healing we find in our relationships is also anything but ordinary. I don't know how to heal a break in my relationships any more than I know how to heal a broken bone, but thankfully, forgiveness is built into our being. We can stumble in our closest relationships, disappointing our partner or our parents or our kids, and still we find our way back to one another. My wife and kids have seen the worst parts of me and I've seen the worst of them, and yet our love for each other is unquestioned. "Love covers a multitude of sins."[9]

Even the mysterious existence of life itself and the universe that sustains it are anything but ordinary. I was missing so much of life when I denied the wonders of everyday experience. I'd been looking to the heavens for what was right here on Earth; it was like finding my lost glasses on top of my head.

When we don't trust that we're loved, we miss the good things that fill our lives, like a child who is unsure of her parents' love and so doesn't see the ways they care for her. Once our eyes are opened to the love that surrounds us, we start to notice that we're cared for in more ways than we knew. We come to realize that the universe is on our side, offering us everything we need, which reinforces the belief in our lovability (see figure 9). We must be lovable if we're constantly taken care of.

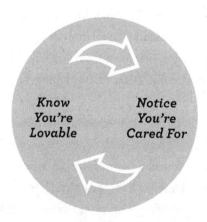

Figure 9

Everything you experience is for you: the patch of sunlight on the wall, the breeze in the air, every color you see. That's why you're witnessing them. You can rest in this assurance, knowing that your ordinary life is filled with good things. So come home to this moment. Come home to where you're sitting. Come home to the people in your life. Come home to who you are. Your extraordinary life is found in everything you already have.

THE OTHER SIDE

A few years ago, it felt as if my life were ending. It wasn't an airplane that was breaking apart, like in my dream, but the structure of my life. I'd lost my physical and mental health, my savings, my connection to friends. I kept bracing for the moment of impact, anticipating a violent crash that would extinguish my life. Whether or not I actually died, I couldn't envision a life after so much loss.

But the death that came was not what I expected. Not only did life continue, it got better. Even before my health improved, I rediscovered a long-abandoned spiritual connection that has carried me through the challenges. Just as in my dream, the death of my old self revealed an eternal connection to everything I love. There was no place left for fear.

So much has changed since that night on the couch. On the other side of pain and loss, I found deep peace. Peace began when I discovered my spirit and found a compassionate witness to my struggles, one that connected me to a larger spirit. I realized I wasn't alone and hadn't been forgotten. The cognitive and behavioral work that followed was built on that firm foundation of mindful awareness. My mind has been renewed. My actions feel integrated with an abiding sense of purpose. I know I'm okay just as I am, so I no longer drive myself mercilessly to do more or live in a constant state of stress.

The integrated approach of Think Act Be that emerged from the intersection of my trials and my training has been discovered by innumerable people across the millennia. Thomas Merton captured it beautifully in *New Seeds of Contemplation*: "The 'spiritual life' is then the perfectly balanced life in which the body with its passions and instincts, the mind with its reasoning and its obedience to principle and the spirit with its passive illumination by the Light and Love of God form one complete [person] who is in God and with God and from God and for God."[10]

We are mind, body, and spirit, and the answers we seek to life's persistent problems will inevitably include tending to our thoughts, acting in alignment with our goals, and being present to our experience.

Whether you're healing from loss or trauma, seeking greater purpose, or struggling with a personal crisis of your own, begin by cultivating presence. Listen for the voice of your spirit, which will be your constant companion. Respond to the spirit's call, and practice right thought and right action. I hope that you find new life, as I have, through the practices of mindful cognitive behavioral therapy. The path forward won't be easy, but it will always be as simple as three words: Think Act Be.

Keep breathing. Mind your thoughts. Do what needs to be done. You're already home.

ACKNOWLEDGMENTS

I'm thankful to so many people who influenced the writing of this book, starting with my parents, Charles and Carolyn Gillihan. Mom, your example of love and care toward the people around you taught me things that didn't fully register until I was much older. Dad, I thought of you a lot during my illness and depression, and wondered what it took for you to keep serving your congregation and your family through your own major struggles. Thank you to both of you for making home a steady and loving force in our lives, despite the many changes of address. I'm grateful I have had the chance to know you better in recent years.

Yonder, Malachi, Timothy, and Charlie, you guys are like brothers to me. Thank you for the commiseration and good humor you've offered as we each found our way in the world. I just wish we lived closer to each other! I long for more time with you. To my parents-in-law, Lance and Cynthia Leithauser, thank you for your love and support for more than twenty-five years.

To my agent Giles Anderson, you have a gift for instantly hearing what I'm trying to say before I've really said it. Thank you for your excitement about this book from the first brief description I gave you at Philo Ridge Farm. That's still the best salad I've ever had.

To my editor Mickey Maudlin, you helped me find the structure to unite a series of ideas into something coherent. Thank you for your patience and insights as you guided me toward writing a better book. Heartfelt thanks to assistant editor Chantal Tom, as well as the rest of the fantastic HarperOne team for your excellent work: Laina Adler, Louise Braverman, Leah Carlson-Stanisic, Yvonne Chan, Ann Edwards, Pat Harris, Amy Reeve, and Amy Sather.

Kim Richardson, you helped me find more of my voice as a writer through our years of collaboration at WebMD. Aria Campbell-Danesh, you're a dearly loved friend and confidant, and I've been buoyed countless times by our inspiring calls across the ocean. Those conversations worked their way into this book. Uh huh. Ray Pasi, I've cherished your friendship, wisdom, encouragement, and sense of humor for more than two decades.

Alice Boyes, I'm grateful to have connected with you through our writing. Thank you for being a supportive friend and fellow author, for introducing me to Giles, and for always seeing through the eyes of abundance. Joel Minden, I feel so fortunate that we found each other, and look forward to many more rewarding conversations with you. Someday we really will meet in person! Paula Ruckenstein, your instruction in those early morning classes at Penn showed me the depth of awareness that's available through yoga.

Rich Mullins, kid brother of St. Frank and fellow Hoosier—I'm sorry we never met in this life, though I feel like I know you through your recorded music and interviews. Your words continue to spread truth and life.

Thank you to the many doctors and other clinicians who did their best to help me figure out what was causing my symptoms, and to the five speech therapists who helped me get my voice back. Special thanks to Diane Gaary for continually surprising me with the breadth of your knowledge and insight, and to clinical nutritionist Josh Gitalis for your lifegiving blend of knowledge and compassion.

My understanding of CBT was shaped by many people, including Alan Goldstein, Robert DeRubeis, Judith Beck, and Aaron T. Beck (who passed away during the writing of this book); my postdoctoral supervisor Elyssa Kushner, who introduced me to mindfulness-based behavior therapy; sleep specialists Michael Perlis and Donn Posner; the late Chris Erickson, who encouraged me to focus on CBT during my doctoral training; and other colleagues too numerous to name, including my #CBTWorks Twitter community. My initial interpretation of CBT was limited by my ability to understand the deeper elements of it, which reflected my own limitations and not those of my instructors.

I am certainly not the first person to bring together mindfulness and CBT. Thank you to the those who influenced my thinking and integration, particularly Steven Hayes, Lizabeth Roemer, Susan Orsillo, and Zindel Segal.

The ideas I express in this book were influenced by conversations I've had with so many individuals on the *Think Act Be* podcast. I appreciate each one of you. To my patients: It has been a privilege to sit with you and to witness your courage—therapy is hard! Thank you for trusting me in this deeply human work.

Lucas, Ada, and Faye, you help me laugh so much every single day, and not just when I laugh at the jokes I tell you. Sometimes you make me laugh even when you're not around because I remember something funny you said or did, or I just remember how much I love the exactly-you-ness of each of you. Thank you for your sweetness and understanding even when my health struggles have affected you. I'm so happy that I get to be your dad!

And finally, to my wife and friend, Marcia Leithauser: this book bears more of your mark than you may know; on so many of our morning walks throughout the pandemic, you offered elegant solutions to the writing dilemmas I was facing. I'm grateful I had the good sense to pursue you in Strasbourg back in 1995, when being with you felt just like coming home. Each time I really see the life we have—a life that's much less perfect and way more beautiful than I had hoped for—I can hardly believe my good fortune. Thank you for holding my hand (and my pain) on all those nights when I had just had it. I've tried to express in these pages how much you have helped me through my illness, but words can't convey all you've done for me or how thankful I am for you.

NOTES

EPIGRAPH

1. Mark S. Burrows and Jon M. Sweeney, *Meister Eckhart's Book of Secrets: Meditations on Letting Go and Finding True Freedom* (Charlottesville, VA: Hampton Roads, 2019), 194.

CHAPTER 1: HEAR THE CALL

1. Names and identifying details of patients have been changed to protect their identity.

CHAPTER 2: CONNECT WITH YOURSELF

1. Adapted from Seth J. Gillihan, *The CBT Deck for Clients and Therapists: 101 Practices to Improve Thoughts, Be in the Moment, and Take Action in Your Life* (Eau Claire, WI: PESI, 2019).
2. Ilia Delio, *The Hours of the Universe: Reflections on God, Science, and the Human Journey* (Maryknoll, NY: Orbis Books, 2021), xv.
3. Omid Naim, "Telling a Better Story about Health and Healing," interview by Seth Gillihan, *Think Act Be*, podcast audio, February 20, 2019, episode 30, https://sethgillihan.com/ep-30-dr-omid-naim/.

CHAPTER 3: FIND LEVERAGE

1. M. Muraven and R. F. Baumeister, "Self-Regulation and Depletion of Limited Resources: Does Self-Control Resemble a Muscle?," *Psychological Bulletin* 126, no. 2 (2000): 247–259, https://doi.org/10.1037/0033-2909.126.2.247.

2. Marcus Aurelius, *Meditations*, trans. Gregory Hays (New York: Modern Library, 2002), 77.

3. David Steindl-Rast and Sharon Lebell, *Music of Silence: A Sacred Journey Through the Hours of the Day* (Berkeley, CA: Ulysses Press, 1998), 5.

4. Most notable among the developers of cognitive therapy were psychiatrist Aaron T. Beck and psychologist Albert Ellis.

5. Epictetus, *The Enchiridion*, in *Discourses and Selected Writings*, ed. and trans. Robert Dobbin (New York: Penguin Books, 2008), 223.

6. This research was conducted, for example, by Ivan Pavlov, Edward Thorndike, and B. F. Skinner.

7. One prominent developer of behavior therapy was the South African psychiatrist Joseph Wolpe.

8. Aurelius, *Meditations*, 119.

CHAPTER 4: SAY YES

1. Steven C. Hayes, "You Want to Feel All of It," interview by Seth Gillihan, *Think Act Be*, podcast audio, October 7, 2020, episode 108, https://sethgillihan.com/ep-108 -dr-steven-c-hayes-you-want-to-feel-all-of-it/.

2. Epictetus, *The Enchiridion*, in *Discourses and Selected Writings*, ed. and trans. Robert Dobbin (New York: Penguin Books, 2008), 222.

CHAPTER 5: PRACTICE MINDFUL AWARENESS

1. Visit https://sethgillihan.com/guided-meditations/ for simple guided meditations.

2. Letícia Ribeiro, Rachel M. Atchley, and Barry S. Oken, "Adherence to Practice of Mindfulness in Novice Meditators: Practices Chosen, Amount of Time Practiced, and Long-Term Effects Following a Mindfulness-Based Intervention," *Mindfulness* 9 (2018): 401–411, https://doi.org/10.1007/s12671-017-0781-3.

3. Chögyam Trungpa, *Shambhala: The Sacred Path of the Warrior* (Boston: Shambhala, 1984).

4. The Raisin Exercise was developed by Jon Kabat-Zinn, creator of the Mindfulness-Based Stress Reduction program.

5. Sam Harris, *Waking Up: A Guide to Spirituality Without Religion* (New York: Simon & Schuster, 2014), 6–7.

6. William James, *The Varieties of Religious Experience: A Study in Human Nature* (Mineola, NY: Dover Publications, 2018), 398–400.

CHAPTER 6: CONNECT WITH YOUR WORLD

1. Isa. 55:2, NIV.

2. Holly B. Shakya and Nicholas A. Christakis, "Association of Facebook Use with

Compromised Well-Being: A Longitudinal Study," *American Journal of Epidemiology* 185, no. 3 (February 1, 2017): 103–211, https://doi.org/10.1093/aje/kww189.

3. D. Nutsford, A. L. Pearson, and S. Kingham, "An Ecological Study Investigating the Association Between Access to Urban Green Space and Mental Health," *Public Health* 127, no. 11 (November 2013): 1005–1011, https://doi.org/10.1016/j.puhe.2013.08.016; Hannah Cohen-Cline, Eric Turkheimer, and Glen E. Duncan, "Access to Green Space, Physical Activity, and Mental Health: A Twin Study," *Journal of Epidemiology & Community Health* 69, no. 6 (June 2015): 523–529, https://dx.doi.org/10.1136/jech-2014-204667.

4. Agnes E. van den Berg et al., "Green Space as a Buffer Between Stressful Life Events and Health," *Social Science & Medicine* 70, no. 8 (April 2010): 1203–1210, https://doi.org/10.1016/j.socscimed.2010.01.002.

5. Caoimhe Twohig-Bennett and Andy Jones, "The Health Benefits of the Great Outdoors: A Systematic Review and Meta-Analysis of Greenspace Exposure and Health Outcomes," *Environmental Research* 166 (October 2018): 628–637, https://doi.org/10.1016/j.envres.2018.06.030.

6. Adapted from Seth J. Gillihan, *The CBT Deck for Anxiety, Rumination, and Worry: 108 Practices to Calm the Mind, Soothe the Nervous System, and Live Your Life to the Fullest* (Eau Claire, WI: PESI, 2020).

7. Mary Oliver, "To Begin With, the Sweet Grass," in *Devotions: The Selected Poems of Mary Oliver* (New York: Penguin Press, 2017), 77.

8. Robin Wall Kimmerer, *Braiding Sweetgrass: Indigenous Wisdom, Scientific Knowledge, and the Teachings of Plants* (Minneapolis, MN: Milkweed Editions, 2020), 118.

9. Lauren E. Sherman et al., "What the Brain 'Likes': Neural Correlates of Providing Feedback on Social Media," *Social Cognitive and Affective Neuroscience* 13, no. 7 (July 2018): 699–707, https://doi.org/10.1093/scan/nsy051.

10. Jon D. Elhai et al., "Problematic Smartphone Use: A Conceptual Overview and Systematic Review of Relations with Anxiety and Depression Psychopathology," *Journal of Affective Disorders* 207 (January 1, 2017): 251–259, https://doi.org/10.1016/j.jad.2016.08.030.

CHAPTER 7: OFFER THANKS

1. "Johnson Oatman," Prabook, accessed 5/25/2022, https://prabook.com/web/johnson.oatman/3767739 Lyric source: https://hymnary.org/text/when_upon_lifes_billows_you_are_tempest.

2. Brenda H. O'Connell, Deirdre O'Shea, and Stephen Gallagher, "Feeling Thanks and Saying Thanks: A Randomized Controlled Trial Examining If and How Socially Oriented Gratitude Journals Work," *Journal of Clinical Psychology* 73, no. 10 (October 2017): 1280–1300, https://doi.org/10.1002/jclp.22469.

3. Joshua A. Rash, M. Kyle Matsuba, and Kenneth M. Prkachin, "Gratitude and Well-Being: Who Benefits the Most from a Gratitude Intervention?," *Applied Psy-*

chology: Health and Well-Being 3, no. 3 (November 2011): 350–369, https://doi
.org/10.1111/j.1758-0854.2011.01058.x.

4. Steven M. Toepfer, Kelly Cichy, and Patti Peters, "Letters of Gratitude: Further Evidence for Author Benefits," *Journal of Happiness Studies* 13 (2012): 187–201, https://doi.org/10.1007/s10902-011-9257-7.

5. Sara B. Algoe, Jonathan Haidt, and Shelly L. Gable, "Beyond Reciprocity: Gratitude and Relationships in Everyday Life," *Emotion* 8, no. 3 (2008): 425–429, https://doi.apa.org/doi/10.1037/1528-3542.8.3.425; Nathaniel M. Lambert et al., "Benefits of Expressing Gratitude: Expressing Gratitude to a Partner Changes One's View of the Relationship," *Psychological Science* 21, no. 4 (2010): 574–580, https://doi.org/10.1177/0956797610364003.

6. William Ferraiolo, *Meditations on Self-Discipline and Failure: Stoic Exercise for Mental Fitness* (Winchester, UK: O-Books, 2017), 163.

7. See https://www.youtube.com/watch?v=BSxPWpLPN7A.

8. Adapted from Seth J. Gillihan, *The CBT Deck for Clients and Therapists: 101 Practices to Improve Thoughts, Be in the Moment, and Take Action in Your Life* (Eau Claire, WI: PESI, 2019).

9. Adapted from Seth J. Gillihan, *The CBT Deck for Anxiety, Rumination, and Worry: 108 Practices to Calm the Mind, Soothe the Nervous System, and Live Your Life to the Fullest* (Eau Claire, WI: PESI, 2020).

10. David Steindl-Rast and Sharon Lebell, *Music of Silence: A Sacred Journey Through the Hours of the Day* (Berkeley, CA: Ulysses Press, 1998), 34.

11. Steindl-Rast and Lebell, *Music of Silence*, 31.

12. See, e.g., O'Connell, O'Shea, and Gallagher, "Feeling Thanks and Saying Thanks."

13. O'Connell, O'Shea, and Gallagher, "Feeling Thanks and Saying Thanks."

14. Steindl-Rast and Lebell, *Music of Silence*, 33.

15. Phil. 4:12, NIV.

16. 1 Thess. 5:18, NIV.

17. James 1:2–3, NIV.

18. Ps. 89:1, NIV.

19. Ps. 13:1, NIV.

20. Johnson Oatman Jr., "Count Your Blessings." Lyric source: https://hymnary.org /text/when_upon_lifes_billows_you_are_tempest

CHAPTER 8: FIND REST

1. David Steindl-Rast, *Gratefulness, the Heart of Prayer: An Approach to Life in Fullness* (New York: Paulist Press, 1984), 181.

2. Jessica de Bloom et al., "Effects of Vacation from Work on Health and Well-Being: Lots of Fun, Quickly Gone," *Work & Stress* 24, no. 2 (2010): 196–216, https://dx.doi.org/10.1080/02678373.2010.493385.

3. Jessica de Bloom, Sabine A. E. Geurts, and Michiel A. J. Kompier, "Vacation (Af-ter-) Effects on Employee Health and Well-Being, and the Role of Vacation Ac-tivities, Experiences, and Sleep," *Journal of Happiness Studies* 14 (2013): 613–633, https://doi.org/10.1007/s10902-012-9345-3.

4. Adapted from Seth J. Gillihan, *The CBT Deck for Clients and Therapists: 101 Prac-tices to Improve Thoughts, Be in the Moment, and Take Action in Your Life* (Eau Claire, WI: PESI, 2019).

5. Marjaana Sianoja et al., "Enhancing Daily Well-Being at Work Through Lunch-time Park Walks and Relaxation Exercises: Recovery Experiences as Mediators," *Journal of Occupational Health Psychology* 23, no. 3 (2018): 428–442, https://doi.org/10.1037/ocp0000083.

6. Thomas Merton, *New Seeds of Contemplation* (New York: New Directions, 2007), 81.

CHAPTER 9: LOVE YOUR BODY

1. Elisabeth Hertenstein et al., "Insomnia as a Predictor of Mental Disorders: A Systematic Review and Meta-Analysis," *Sleep Medicine Reviews* 43 (February 2019): 96–105, https://doi.org/10.1016/j.smrv.2018.10.006.

2. M. Daley et al., "Insomnia and Its Relationship to Health-Care Utilization, Work Absenteeism, Productivity, and Accidents," *Sleep Medicine* 10, no. 4 (April 2009): 427–438, https://doi.org/10.1016/j.sleep.2008.04.005.

3. Wendy M. Troxel et al., "Marital Quality and the Marital Bed: Examining the Co-variation Between Relationship Quality and Sleep," *Sleep Medicine Reviews* 11, no. 5 (October 2007): 389–404, https://doi.org/10.1016/j.smrv.2007.05.002.

4. Richard Rohr, "Living Fully," Center for Action and Contemplation, April 3, 2019, https://cac.org/living-fully-2019-04-03/.

5. Jason C. Ong et al., "A Randomized Controlled Trial of Mindfulness Meditation for Chronic Insomnia," *Sleep* 37, no. 9 (September 1, 2014): 1553–1563, https://doi.org/10.5665/sleep.4010.

6. M. Alexandra Kredlow et al., "The Effects of Physical Activity on Sleep: A Meta-Analytic Review," *Journal of Behavioral Medicine* 38 (2015): 427–449, https://doi.org/10.1007/s10865-015-9617-6.

7. Ana Kovacevic et al., "The Effect of Resistance Exercise on Sleep: A Systematic Review of Randomized Controlled Trials," *Sleep Medicine Reviews* 39 (June 2018): 52–68, https://doi.org/10.1016/j.smrv.2017.07.002.

8. Kredlow et al., "Effects of Physical Activity on Sleep."

9. Eric Suni, "Sleep Hygiene: What It Is, Why It Matters, and How to Revamp Your Habits to Get Better Nightly Sleep," Sleep Foundation, updated November 29, 2021, https://www.sleepfoundation.org/articles/sleep-hygiene.

10. Saundra Dalton-Smith, *Sacred Rest: Recover Your Life, Renew Your Energy, Restore Your Sanity* (New York: FaithWords, 2017), 8.

11. I'm grateful to Rebecca Stoltzfus, president of Goshen College, for helping to open my eyes to the sacred nature of sleep.

12. William James, *The Varieties of Religious Experience: A Study in Human Nature* (Mineola, NY: Dover Publications, 2018), 277.

13. Gen. 41, NIV.

14. Matt. 2:12, NIV.

15. Ps. 127:2, NIV.

16. Swami Krishnananda, "Consciousness and Sleep," in *The Māndūkya Upanishad* (Rishikesh, India: Divine Life Society, 1996), https://www.swami-krishnananda .org/mand/mand_5.html.

17. Hazrat Inayat Khan, "Sufi Teachings: The Mystery of Sleep," Hazrat Inayat Khan Study Database, http://hazrat-inayat-khan.org/php/views.php?h1=30&h2=33.

18. Bhante Shravasti Dhammika, "Sleep," Guide to Buddhism A to Z, https://www .buddhisma2z.com/content.php?id=385.

19. Bhante Shravasti Dhammika, "Dreams," Guide to Buddhism A to Z, https:// www.buddhisma2z.com/content.php?id=116.

20. Nechoma Greisman, "The Philosophy of Sleep," Chabad.org, https://www.chabad .org/library/article_cdo/aid/97560/jewish/The-Philosophy-of-Sleep.htm.

21. Gary M. Cooney et al., "Exercise for Depression," *Cochrane Database of Systematic Reviews*, no. 9, art. CD004366 (September 12, 2013), https://doi .org/10.1002/14651858.CD004366.pub6.

22. Felipe B. Schuch et al., "Physical Activity Protects from Incident Anxiety: A Meta-Analysis of Prospective Cohort Studies," *Depression & Anxiety* 36, no. 9 (September 2019): 846–858, https://doi.org/10.1002/da.22915.

23. Amanda L. Rebar et al., "A Meta-Meta-Analysis of the Effect of Physical Activity on Depression and Anxiety in Non-clinical Adult Populations," *Health Psychology Review* 9, no. 3 (2015): 366–378, https://doi.org/10.1080/17437199.2015.1022901.

24. Charles B. Eaton et al., "Cross-Sectional Relationship Between Diet and Physical Activity in Two Southeastern New England Communities," *American Journal of Preventive Medicine* 11, no. 4 (July–August 1995): 238–244, https://doi .org/10.1016/S0749-3797(18)30452-5.

25. Jacobo Á. Rubio-Arias et al., "Effect of Exercise on Sleep Quality and Insomnia in Middle-Aged Women: A Systematic Review and Meta-Analysis of Randomized Controlled Trials," *Maturitas* 100 (June 1, 2017): 49–56, https://doi .org/10.1016/j.maturitas.2017.04.003.

26. Richard M. Ryan and Edward L. Deci, "Self-Determination Theory and the Facilitation of Intrinsic Motivation, Social Development, and Well-being," *American Psychologist* 55, no. 1 (2000): 68–78, https://doi.apa.org/doi/10.1037/0003-066X.55.1.68.

27. Adapted from Seth J. Gillihan, *The CBT Deck for Anxiety, Rumination, and Worry: 108 Practices to Calm the Mind, Soothe the Nervous System, and Live Your Life to the Fullest* (Eau Claire, WI: PESI, 2020).

28. The "SMILES" trial and other research studies have shown that improvements in diet can relieve depression and anxiety and that vitamin and mineral supplements can improve recovery from traumatic events. See, e.g., Felice N. Jacka et al., "A Randomised Controlled Trial of Dietary Improvement for Adults with Major Depression (the 'SMILES' Trial)," *BMC Medicine* 15, no. 23 (2017): https://doi.org/10.1186/s12916-017-0791-y.

29. My understanding of many of the nutrition-related issues in this chapter has been enriched through conversations with my friend and fellow psychologist Aria Campbell-Danesh.

30. Try Ina Garten's recipe: https://www.foodnetwork.com/recipes/ina-garten/roasted-brussels-sprouts-recipe2-1941953.

31. Adapted from Seth J. Gillihan, *The CBT Deck for Clients and Therapists: 101 Practices to Improve Thoughts, Be in the Moment, and Take Action in Your Life* (Eau Claire, WI: PESI, 2019).

CHAPTER 10: LOVE OTHERS

1. Edward Brodkin and Ashley Pallathra, *Missing Each Other: How to Cultivate Meaningful Connections* (New York: PublicAffairs, 2021), 2.

2. Adapted from Seth J. Gillihan, *The CBT Deck for Clients and Therapists: 101 Practices to Improve Thoughts, Be in the Moment, and Take Action in Your Life* (Eau Claire, WI: PESI, 2019).

3. Adapted from Gillihan, *CBT Deck for Clients and Therapists*.

4. Ross Gay, *The Book of Delights* (Chapel Hill, NC: Algonquin Books of Chapel Hill, 2019), 97.

5. Jared Byas, *Love Matters More: How Fighting to Be Right Keeps Us from Loving Like Jesus* (Grand Rapids, MI: Zondervan, 2020), 69.

CHAPTER 11: WORK IN ALIGNMENT

1. Kahlil Gibran, *The Prophet* (Hertfordshire, UK: Wordsworth Editions, 1997), 13.

2. Edward L. Deci and Richard M. Ryan, "Self-Determination Theory: A Macrotheory of Human Motivation, Development, and Health," *Canadian Psychology* 49, no. 3 (2008): 182–185, https://doi.org/10.1037/a0012801.

3. Anja Van den Broeck et al., "Capturing Autonomy, Competence, and Relatedness at Work: Construction and Initial Validation of the Work-Related Basic Need Satisfaction Scale," *Occupational and Organizational Psychology* 83, no. 4 (December 2010): 981–1002, https://doi.org/10.1348/096317909X481382.

4. Marie-Hélène Véronneau, Richard F. Koestner, and John R. Z. Abela, "Intrinsic Need Satisfaction and Well-being in Children and Adolescents: An Application of the Self-Determination Theory," *Journal of Social and Clinical Psychology* 24, no. 2 (July 2005), https://doi.org/10.1521/jscp.24.2.280.62277; José A. Tapia

Granados et al., "Cardiovascular Risk Factors, Depression, and Alcohol Consumption During Joblessness and During Recessions Among Young Adults in CARDIA," *American Journal of Epidemiology* 187, no. 11 (November 2018): 2339–2345, https://doi.org/10.1093/aje/kwy127; Noortje Kloos et al., "Longitudinal Associations of Autonomy, Relatedness, and Competence with the Well-being of Nursing Home Residents," *Gerontologist* 59, no. 4 (August 2019): 635–643, https://doi.org/10.1093/geront/gny005.

5. Adapted from Seth J. Gillihan, *The CBT Deck for Clients and Therapists: 101 Practices to Improve Thoughts, Be in the Moment, and Take Action in Your Life* (Eau Claire, WI: PESI, 2019).

6. Jessica de Bloom, Sabine A. E. Geurts, and Michiel A. J. Kompier, "Vacation (After-) Effects on Employee Health and Well-Being, and the Role of Vacation Activities, Experiences, and Sleep," *Journal of Happiness Studies* 14 (2013): 613–633, https://doi.org/10.1007/s10902-012-9345-3.

7. Gibran, *The Prophet*, 13.

8. David K. Reynolds, *A Handbook for Constructive Living* (Honolulu: University of Hawai'i Press, 2002).

9. Adapted from Gillihan, *CBT Deck for Clients and Therapists*.

10. Adapted from Gillihan, *CBT Deck for Clients and Therapists*.

11. Eugene H. Peterson, *The Contemplative Pastor: Returning to the Art of Spiritual Direction* (Grand Rapids, MI: William B. Eerdmans, 1993), 19.

12. Thomas Merton, *New Seeds of Contemplation* (New York: New Directions, 2007), 45.

13. I'm grateful to Gregg Krech for stimulating my thinking in this regard. Gregg Krech, *Naikan: Gratitude, Grace, and the Japanese Art of Self-Reflection* (Berkeley, CA: Stone Bridge Press, 2002).

14. Robin Wall Kimmerer, *Braiding Sweetgrass: Indigenous Wisdom, Scientific Knowledge, and the Teachings of Plants* (Minneapolis, MN: Milkweed Editions, 2020), 232.

15. Merton, *New Seeds of Contemplation*, 19.

16. Stephen Mitchell, trans., *Bhagavad Gita: A New Translation* (New York: Three Rivers Press, 2000), 51, 63.

17. Rom. 12:1, NRSV.

18. Eugene H. Peterson, *The Message: The Bible in Contemporary Language* (Colorado Springs, CO: NavPress, 2018), 942.

19. Gibran, *The Prophet*, 14.

CHAPTER 12: LIVE WITH PURPOSE

1. "'The Three Questions' by Leo Tolstoy," *Plough Quarterly* 7 (Winter 2016), https://www.plough.com/en/topics/culture/short-stories/the-three-questions.

2. David Steindl-Rast and Sharon Lebell, *Music of Silence: A Sacred Journey Through the Hours of the Day* (Berkeley, CA: Ulysses Press, 1998), 8.

3. Adapted from Seth J. Gillihan, *The CBT Deck for Clients and Therapists: 101 Practices to Improve Thoughts, Be in the Moment, and Take Action in Your Life* (Eau Claire, WI: PESI, 2019).

4. Steindl-Rast and Lebell, *Music of Silence*, 10.

5. Frederick Buechner, *Wishful Thinking: A Theological ABC* (New York: Harper & Row, 1973), 95.

6. Thomas Merton, *New Seeds of Contemplation* (New York: New Directions, 2007), 223–224.

7. Adapted from Seth J. Gillihan, *The CBT Deck for Anxiety, Rumination, and Worry: 108 Practices to Calm the Mind, Soothe the Nervous System, and Live Your Life to the Fullest* (Eau Claire, WI: PESI, 2020).

8. David Steindl-Rast, *Gratefulness, the Heart of Prayer: An Approach to Life in Fullness* (New York: Paulist Press, 1984), 40.

CHAPTER 13: COME HOME

1. Rich Mullins, "Growing Young," in *The World as Best as I Remember It*, vol. 2 (Reunion Records, 1991).

2. Rich Mullins, *Live from Studio B* (live television concert with A Ragamuffin Band at LeSEA Broadcasting, South Bend, IN, March 14, 1997), https://www.youtube .com/watch?v=jkPuHReiFeM.

3. Thomas Merton, *New Seeds of Contemplation* (New York: New Directions 2007), 81.

4. Dan. 5:27, NRSV.

5. Mark 8:36, KJV.

6. Mark S. Burrows and Jon M. Sweeney, *Meister Eckhart's Book of Secrets: Meditations on Letting Go and Finding True Freedom* (Charlottesville, VA: Hampton Roads, 2019), 77.

7. Rom. 8:39, NRSV.

8. Adapted from Seth J. Gillihan, *The CBT Deck for Clients and Therapists: 101 Practices to Improve Thoughts, Be in the Moment, and Take Action in Your Life* (Eau Claire, WI: PESI, 2019).

9. 1 Pet. 4:8, NRSV.

10. Merton, *New Seeds of Contemplation*, 140.